T0332258

Ultra Wideband

Demystified

Technologies, Applications, and System Design Considerations

RIVER PUBLISHERS SERIES IN COMMUNICAITONS

Volume 4

Consulting Series Editor

Marina Ruggieri
University of Roma Tor Vergata
Italy

Other books in this series:

Volume 1
4G Mobile & Wireless Communications Technologies
Sofoklis Kyriazakos, Ioannis Soldatos, George Karetsos
September 2008
ISBN: 978-87-92329-02-8

Volume 2
Advances in Broadband Communication and Networks
Johnson I. Agbinya, Oya Sevimli, Sara All, Selvakennedy Selvadurai,
Adel Al-Jumaily, Yonghui Li, Sam Reisenfeld
October 2008
ISBN: 978-87-92329-00-4

Volume 3
*Aerospace Technologies and Applications for Dual Use A New World of
Defense and Commercial in 21st Century Security*
General Pietro Finocchio, Ramjee Prasad, Marina Ruggieri
November 2008
ISBN: 978-87-92329-04-2

Ultra Wideband

Demystified

Technologies, Applications, and System Design Considerations

Sunil Jogi

Manoj Choudhary

Routledge
Taylor & Francis Group

LONDON AND NEW YORK

Published 2009 by River Publishers
River Publishers
Alsbjergvej 10, 9260 Gistrup, Denmark
www.riverpublishers.com

Distributed exclusively by Routledge
4 Park Square, Milton Park, Abingdon, Oxon OX14 4RN
605 Third Avenue, New York, NY 10017, USA

Ultra Wideband Demystified: Technologies, Applications, and System Design Considerations / by Sunil Jogi, Manoj Choudhary.

Routledge is an imprint of the Taylor & Francis Group, an informa business

ISBN 978-87-92329-14-1 (print)

While every effort is made to provide dependable information, the publisher, authors, and editors cannot be held responsible for any errors or omissions.

Foreword

Ultra Wideband (UWB) seeks to unleash a new frontier in very high speed short range wireless connectivity space. With applications in consumer electronics (e.g. connecting a High Definition TV to a DVD player wirelessly), in mobile phone space (e.g. transferring high definition images/files to/from a PC/laptop/printer) and in the PC space (e.g. connecting a PC and printer/camcorder wirelessly), it has quickly emerged as the most wanted "last few meters" wireless connectivity solution.

The event of Federal Communications Commission (FCC) allowing UWB for commercial use in 2002, with maximum limit on radiated power, accelerated the growth of UWB. Despite facing competitive challenges during standardization activities at IEEE, the products based on UWB technology have started to appear in the marketplace and catching the fancy of both technologists and consumers. The recent Consumer Electronics Shows have been both a testimony of maturity and variety of products based on UWB, as well as the excitement amongst the community.

Of particular interest, has been the UWB technology based on the Multi-band OFDM (MB-OFDM). Besides technological superiority, this has the benefits of a larger spectrum of companies supporting the development of specifications at the WiMedia Alliance and products based on those. These attempts received major impetus when MB-OFDM based UWB solution was adopted by ECMA (European Computers Manufacturers Association), USB-IF (Universal Serial Bus Implementers Forum) defining Wireless USB aligned with WiMedia specifications, and Bluetooth SIG inclined to consider this as an alternate radio platform. Regulatory progress in several countries (United States, Japan, China, Europe etc.) has accelerated the technology development toward potential deployments. The MB-OFDM technology incorporating the Detect and Avoid (DAA) mechanisms has gone a long way toward providing flexibility in resolving potential coexistence issues with other technologies in the band of interest.

A few good books have recently been published covering the emergence of UWB technology and foundations of various building blocks of the technology.

This book is very timely, unique and fresh in its approach, coming from engineers who have been involved in system design and standard development stages. In particular, the book stands out amongst other literature available because it highlights system designer's viewpoints and because it covers the whole gamut of technology from practitioner's viewpoints. The book describes the following:

- Trends in wireless and the need for very high speed short range wireless leading to evolution of UWB.
- Perspectives in the journey toward evolution of UWB standards.
- MB-OFDM technology as adopted by ECMA.
- System design considerations at Physical layer and Medium Access Control layer and the Hardware/Software partitioning approaches.
- Notes on regulatory frameworks, security aspects and power management techniques essential to UWB usage in consumer devices like portable handheld mobile devices.
- Adaptations to multiple applications with specific emphasis to Wireless USB and design considerations.
- UWB in the communication ecosystem for the converging marketplace.

I would strongly recommend this book to System Designers, Practicing Engineers, Researchers in Academia and Industry, Product Marketing and Technical Strategists for a comprehensive reading on the emerging UWB technologies. I commend Sunil Jogi and Dr. Manoj Choudhary for a very timely contribution.

— Bart Vertenten
Chief Architect Connectivity, NXP Semiconductors
Chair/Technical Editor, Wireless USB 1.1 Protocol working group

About the Author

Sunil Jogi: Sunil Jogi has been involved in Ultra Wideband standardization since beginning. He has 8 years of experience in wireless semiconductor industry involving implementation, research, system architecture, standardization, prototype development leading to products, product certification and interoperability for short range wireless technologies. Currently, he has been working with NXP Semiconductors for the UWB program. He has been representing NXP Semiconductors in the WiMedia Alliance. Prior to this, he worked with Samsung Electronics in the "Next Generation Wireless" group. At Samsung Electronics, he worked on standardization aspects for IEEE, Multi-band OFDM Alliance (MBOA) and WiMedia Alliance. He has extensively worked on hardware–software co-design of multiple wireless MAC systems such as UWB, Wireless USB, IEEE 802.11a/b, IEEE 802.15.3, IEEE 802.11s etc. and has experience in defining adaptation layer for different connectivity. He has co-authored WiMedia MAC and WLP specifications and also authored/co-authored several patents applications in areas of short range wireless.
Contact: sunil@suniljogi.com

Manoj Choudhary: Manoj Choudhary received his PhD from the Indian Institute of Technology (IIT) Kanpur. Currently, he is working with Texas Instruments at Bangalore (India) as Development Manager and Member Group Technical Staff in the area of modem software development for mobile handsets. Prior to joining TI, he worked with Samsung Electronics as Senior Development Manager and Group Head of "Next Generation Wireless" group focusing on research, standardization and prototype development of emerging wireless technologies such as Ultra Wideband, 3G-WLAN interworking and 4G. He was associated with Ultra Wideband standardization efforts in IEEE and Multi-band OFDM Alliance (MBOA) since beginning. He has been one of the authors of Ultra Wideband MAC specification at the WiMedia Alliance and also served as co-chair of Technical Steering Committee for the WiMedia Alliance during 2005–2006. Manoj has delivered several invited talks and tutorials in national/international conferences/symposiums and industry forums. Manoj has been an author/co-author of several patent applications in US, Korea, European Union and India.

Contact: manojc@gmail.com

Acknowledgments

Toward writing a book on such a recent technology, several persons have contributed their significant time and effort in providing specific inputs and review comments as well as by invaluable interactions during author's participation into this exciting technology. Without their help and guidance, this task would have been almost impossible.

First of all, the authors would like to thank their colleagues and friends, who read the drafts of the book carefully despite their busy schedules, and provided with invaluable comments and suggestions: thanks to Sushil Kumar and Rahul Kulkarni for providing overall comments and technical suggestions for the book; Bart Vertenten for his review; Rakesh Ughreja and Luciano Trunzo for review and suggestions on Wireless USB; Jaykishan Nirmal for assistance on history and security; Surjyakantha Mohapatra for wide area networks; and Vishal Jogi for helping out on front cover. Special thanks are due to Mario Pasquali for providing information about Ellisys UWB tools; and Mike Micheletti for Lecroy UWB tools.

Authors are grateful to their several collaborators and co-workers over the years who have enriched them through deep technical insights. In this regard, they would like to specially mention Balaji Holur, Dr Yongsuk Kim, Kisoo Chang, Prashant Wason, Thenmozhi Arunan, Vivek Kumar, Manu Agarwal, Rakesh Ughreja, Madhanraj, Sundaresan, Devendra Purbiya, Venkatraman, Giriraj Goyal, Kyeong Hur, Rakesh Taori, Dr Junseo Lee, Junhaeng Cho, Arun Naniyat, Dr Anand Santhana Krishna, Dr S. Raghunadan, Sushil Kumar, Rahul Kulkarni, Pandiarajan, S.J. Rahul, Dattatreya, Vamsi Krishna, Rakesh Raman, Manoj Kamath, Harish Vaidya, Sachin Athalye, Bart Vertenten, Luciano Trunzo, Jo Degraef, Koen Derom, Alexandre Esquenet, Bruno Rahir, Lixin Liang, Dave Sroka, Tom Sun, Stefan Janssens, Kawshol Sharma, Ankur Goel, Shakti Amarendra, Visweswara Rao, Krishna Chaitanya, Krishan Singh,

Jiang Qian, Albert Goh, Phani Kumar, Sasidhar Ravella, Amit Jhawar, Dr P.K. Chatterjee, Dr V. Sinha, Dr A.K. Chaturvedi, Dr Joseph John, Dr Atul, Dr Alpana, M.D. Umrania, P.B. Rawal, Hunju Park, Gyu Chool Kim, Karthik Chandrasekhar, R. Rajendran, and S. Nagarajan.

Authors owe gratitude to their respective employers for permitting them to work on this book. They would like to thank P. Jayalakshmi, S.J. Rahul, Dr N.S. Murty, Rajeev Mehtani, Giten Kulkarni, Bruno Rahir, Maarten Nollen, Olivier Potin, and Balaji Holur.

Authors would also like to thank the participants of MB-OFDM Alliance, WiMedia Alliance, and IEEE work groups, who helped them to enrich the knowledge of short range wireless technologies and specially Ultra Wideband.

Special thanks to the team of River Publishers, especially Elisabeth Jansen for guidance and keeping authors on schedule for the project.

Personally, friends and family have provided support and encouragement for which authors are deeply grateful. Sunil is thankful to his wife Jalpa for continuous encouragement and support for the project and for walking together in the journey of life. Manoj would like to acknowledge his heartfelt thanks to his wife Bhawana and their lovely daughter Snigdha for making life so beautiful with their presence. Authors remain indebted to their parents and teachers who gave them life, intellectual strength and inspiration.

Authors

Preface

As the world moves aggressively toward "removing" wire everywhere, there is a need to have a very high speed short range wireless system catering to last few meters (akin to the last mile approaches in telecommunications). This reflects a magnitude change in data rates from very common Bluetooth enabled mobile handsets existing today that allow peer-to-peer network formation for data sharing. For example, one may like to have a computer and a printer connected wirelessly. Or, a TV and a DVD player or Set-top box connected wirelessly. Or, transferring high definition video to High Definition TV (HDTV) from the media player. Combination of such use cases in short range networks require very high data rates (exceeding 100 Mbps) to be supported for multiple devices simultaneously with limited mobility over smaller distances (typically 10 m or less).

Ultra Wideband (UWB) has come a long way from being used in strategic military and radar applications. Ever since the Federal Communications Commission (FCC) allowed in 2002 the commercial usage of UWB in the unlicensed bands, there has been tremendous interest in the several industry segments (Semiconductor, Mobile handset, Consumer electronics, PC makers) to define standards so that interoperable products can be built to take advantage of immense market opportunity. Today, standards have evolved that allows data rates up to 480 Mbps over a distance of 3 m and up to 110 Mbps over a distance of 10 m. Industry is on the threshold of bringing products based on this radical technology to the consumers that is forever going to change the way we communicate and live. It enables unwiring high speed wired connections and discovers many new applications for high speed multimedia content transfer using UWB.

This book focuses on emergence of UWB, evolving standards, practical system design/implementation, and real life applications. The book also covers the standards development efforts, regulatory constraints, security consider-ations, and power management techniques essential to its usage in portable handheld mobile devices. This book also covers UWB radio platform usage for adapting different existing upper layer protocols and tools for enabling existing applications in wired domain, new upper layer adaptation and new application development. This book is not intended to be a replacement for UWB standards.

The content of the book is prepared to be useful to system designers, practicing engineers, researchers and academicians and in general to anyone interested in the exciting world of UWB and of very high speed short range wireless. The readers with basic background of wireless communications and a preliminary understanding of embedded systems can use it for UWB system design, usage, and applications. Practicing engineers could be in any job func-tions like software design engineer, system design engineer, firmware engi-neer, scientists, researchers, strategists, those involved in regulatory policies and standardization activities, who can use this book as a reference on UWB systems.

Authors

Contents

Table of Figures

Table of Tables

1

Introduction to Short Range Wireless

1.1 Trends in Wireless

Wireless has become part and parcel of our day-to-day lifestyle. Desire to be free from the shackles of wires and ease of connectivity is fueling this growth in all aspects to create a truly "wireless world." Mobile phones, PDAs, laptop computers are overtaking traditional desktop computers and fixed wired line phones in popularity due to ease of use and flexibility of movement without the mess of wires. Capabilities of devices are increasingly being extended, moving toward "all in one" devices. One uses mobile phone for making calls, receiving messages on pager, checks email on PDA, and uses internet over laptop computer, all through wireless. Emergence of various wireless technologies is enabling "unwiring" of many applications in the business and home environment. Newer applications are emerging to take advantage of emergence of new technologies.

Cellular technologies have been the most recognizable and popular of the wireless technologies. It has allowed people to be "connected" while on move. Second generation of wireless technologies (GSM/GPRS) is now prevalent in cell phones across the globe. 3G has been around last few years. As time goes, application requirements change, and eventually faster and cheaper technology emerge with increasing new service base. As regulatory bodies consider opening many new frequency bands to combat spectrum scarcity, research is picking up on using new frequency bands and techniques for wireless communication that evolve around using the precious spectrum efficiently. Different frequency bands and modulation schemes are being proposed for increasing data rates. Research in next generation cellular technologies like 3.5G and 4G and others (e.g. WiMAX) are gathering up the momentum. There have been claims that these new technologies are cheaper and faster than 3G besides providing opportunities for many more services and applications.

Trial runs and deployments of next generation mobile wireless networks are going on in all over the world. It is expected to be used not only for mobile phones but many other gadgets, PDAs, and laptop computers. Not merely for voice and low data service but it is expected to have streaming capabilities and multimedia content download services with many other new applications.

Chip manufacturers are busy with creating new chips for base stations and end devices to cross new frontiers of new cellular technologies. Combo chips, which provide multiple wireless connectivity technologies in single chip, are also being available in the marketplace. Interworking of different wireless technologies is being targeted for seeking maximum advantage from peculiarities of each wireless technology to create a seamless and enriching experience for the end user.

The Global Positioning System (GPS) and tracking applications are also becoming usual nowadays. One could see most of the cars and buses being fitted with GPS devices. There are GPS enabled cell phones, navigating devices used in vehicles, and vehicle tracking devices becoming common these days.

A look at our home and office reveals the need for short range wireless (10 m to 100 m range) with high data rate and less power consumption. Large file download, live media streaming, infrastructure networking, quick *ad-hoc* networking are expected applications of today's short range wireless. Improvements and innovations to provide such applications for short range wireless are one of the most debated and discussed topics in technology circles.

Wireless Local Area Network (WLAN) technologies have enabled the mobility in the laptop world in office and home environments. While WLAN based on IEEE 802.11 a/b/g are popular and are getting more efficient, IEEE 802.11n is emerging as a new WLAN technology to provide much higher throughputs. IEEE 802.11n is based on exploiting advantages of the Multiple Input Multiple Output (MIMO) technology to provide much higher throughput, support of Quality of Service (Qos), and support of isochronous traffic. Short range wireless with these features enables many new kinds of usage and can be a replacement of many existing wired technologies. Research is also going on for sharing radio for multiple protocols.

Another embryonic class of wireless is of wearable devices and communication amongst those devices. Previously used in healthcare domain, this is now expanding application base by including sports, military, and security domains. One more emerging field is sensor networking, mainly used in application like monitoring, object tracking, fire detection, traffic monitoring, and healthcare. This may incorporate many different communication technologies of short range wireless or wide area networks depending on deployment and physical location. Extensive research is still going on in this area so that the technology can support home automation and security with smart body networks.

New generation of wireless technologies are providing higher data rates, consuming lower power, have lower cost and huge application capabilities, but security is still considered to be a concern. Though wireless unwires the wired connectivity, because of open air medium wireless does not provide security similar to closed wired environment and is prone to hackers and man-in-middle attacks. Wireless security is a very challenging area of research. Many data encryption standards have been evolved and broken by hackers in due course of time. Security infrastructures are provided for wireless entities like access point and wireless router to prevent unauthorized user access. Unified Threat Management (UTM) provides anti-spam, anti-spyware, anti-virus combined with firewall and Virtual Private Network (VPN).

Who and what should go to wireless? Answer is everyone and everything. People at this time are looking for seamless mobility and retaining connection at each and every place, while being in office, while driving, being at home. It is only possible through wireless infrastructure and ecosystem where multiple wireless technologies interwork and provide seamless mobility and hand-off between different wireless connectivity technologies. As a result, continuous and seamless service can be provided to consumer, without the consumer noticing or bothering much about which wireless technology is providing specific service to him at the moment. These are some of the common driving force for all wireless technologies.

1.2 Wireless Technologies Characterization and Standards

In last few years, wireless technologies have advanced a lot to cater to next generation's services and application requirements. Many new technologies and standards have emerged, which provide support of unwiring existing applications while ensuring high throughput and better security. New technologies have also led to unique new use cases resulting in improved user experience.

For increased acceptance and market reach, as well as for interoperability of multiplicity of devices, standardization is much desirable for wireless technologies real field usage and deployment. Institute of Electrical and Electronics Engineers, Inc. (IEEE), the world's leading professional association for the advancement of technology, and many other national as well as international standardization organization and industry alliances are working hard to define standards for wireless technologies. These standards enable the products interoperation and can reliably be used in different environments.

From the radio range perspective wireless technologies can basically be characterized in (a) Wireless Wide or Metropolitan Area Network (WWAN or WMAN) also called Cellular Network, which is of long range of a few kilometers; (b) Wireless Local Area Network (WLAN), which is of short range compared to cellular network, range of around 100 m; and (c) Wireless Personal Area Network (WPAN), which caters to even shorter range for personal networking up to few meters (approximately 10–50 m).

Cellular network is one of the most deployed and used network. It is still evolving for improvements in different aspects of user experience. In early 1980s, it started with analog radio technologies, which evolved to today's 2G and 3G networks and in near future 4G networks. Multiple standardization bodies and industry alliances worked for standardization of cellular network to ensure interoperability and seamless roaming. Initially, North America adopted IS-136 (Time Division Multiple Access — TDMA) and IS-95 (Code Division Multiple Access — CDMA) under Telecommunications Industry Association (TIA). On the other hand, Europe adopted Global System for Mobile communications, more popularly known as GSM technology for cellular network under GSM Association. It is widely accepted in many parts of the world. During the same time, NTT DoCoMo developed its own Personal Digital Cellular (PDC) network for Japan. Later on, GSM was enhanced to General Packet Radio Service (GPRS) to provide packet oriented mobile data services. CDMA was enhanced to CDMA 1xEV-DO and EV-DV.

In year 1998, Third Generation Partnership Project (3GPP) was formed to create next generation wireless technology standards. It is a collaborative effort of groups of telecommunications associations and resulted in evolution of GSM, GPRS, and W-CDMA standards. 3GPP has sub groups working on different architecture and protocols of the system for core network, radio access network, and mobile handsets. On the other hand, same time, 3GPP2 was formed to develop new standard for packet switched network using CDMA2000 again based on CDMA radio technology protocol. Currently, 3GPP is working toward Long Term Evolution (LTE) project to define systems that can provide the data rates in the range of 100 Mbps.

Meanwhile, IEEE's metropolitan area group extended the IEEE 802.16 standard, which provided high speed fixed wireless connectivity, to include provisions for mobile environments. The WMAN standard IEEE 802.16d is for fixed systems, while IEEE 802.16e was developed for including mobility support. WiMAX forum has adopted these standards while defining interoperability and certification criteria, referring these standards as Fixed WiMAX and Mobile WiMAX, respectively. This can provide high data rate up to 70 Mbps.

In local area network, 802.11, wireless replacement of wired Ethernet LAN is one of very successful standard defined by IEEE 802.11 work group. There are different set of standards with variations of 802.11 defined by subgroups of IEEE 802.11. 802.11b is an initial standard with physical data rate up to 11 Mbps over a frequency band of 2.4 GHz. 802.11a provides higher data rate up to 54 Mbps over a frequency band of 5 GHz. Improvements continued over these standards and 802.11g provided 54 Mbps over 2.4 GHz. With few more improvements and additions, 802.11e supported QoS and 802.11i supported enhanced security. Now, 802.11n using MIMO antenna is aiming for more than 100 Mbps data rate. IEEE 802.11s is targeting layer two mesh networking support.

European alternative of 802.11 is HiperLAN, defined by European Telecommunications Standards Institute (ETSI). HiperLAN/1 uses frequency band of 2.4 GHz and data rate around 20 Mbps. HiperLAN/2 is higher data rate solution that provides data rate up to 54 Mbps at 5 GHz with QoS support.

Coming to the short range wireless for personal area networks, many different industry alliances worked on WPAN and developed the standards. Bluetooth and Zigbee are well known examples of that. IEEE being a major standardization body worldwide, created IEEE 802.15 work group for standardization of wireless personal area network. Different groups of 802.15 targeted various applications and data rate for WPAN, including features like Mesh networking. Industry standards, Bluetooth, and Zigbee are adopted by IEEE under 802.15. Near Field Communication (NFC), developed by NFC Forum is adopted by European Computer Manufacturers Association

(ECMA) and International Organization for Standardization (ISO). HomeRF for wireless data sharing between home devices is defined by HomeRF Working Group, a consortium of mobile wireless companies. The Infrared Data Association (IrDA) defines communication standards and protocol standards for the short data exchange over infrared light, for WPANs. Of late, IEEE and other industry bodies developed enough interest for Ultra Wideband (UWB) technologies for very high data rate short range wireless. This book focuses on UWB technologies, standards, and applications.

Different standards bodies and alliances, like IEEE, European Computer Manufacturers Association (ECMA), International Organization for Standardization (ISO), American National Standards Institute (ANSI), played a key role in the development of wireless technology and market adoption of the technology.

1.3   Short Range Wireless

Short Range Wireless communication is quite different compared to well known cellular network kind of wide area communication network. It removes annoying mess of cables around our gadgets in day-to-day usage. It makes connections easy and simple without need of cables or connecters. It is a technology to enable rapid *ad-hoc* connections, and sometimes also automatic and unconscious connections. Short range wireless also enables new kind of applications like monitoring and sensor networking.

Short range wireless is generally classified in different types based on its range and usage.

Popular WLAN technology is a replacement of wired LAN having radio range around 30 m to 100 m. Mainly used in indoor environments, like office network or in hotel, nowadays also used for networking at home and home-office, is sometimes used in outdoor environment also. IEEE 802.11 has been a successful technology for WLAN. It supports *ad-hoc* networking as well as infrastructure networking, which makes it truly a replacement of wired LAN. Signal of 802.11 perpetrates through walls and makes it ideally suited for indoor environment. WLAN is mainly used for internet and local network connectivity. Newer standards support QoS and security requirements for new applications with higher throughput. Figure 1.1 shows Wireless LAN *ad-hoc* and infrastructure network with connection to internet backbone network.

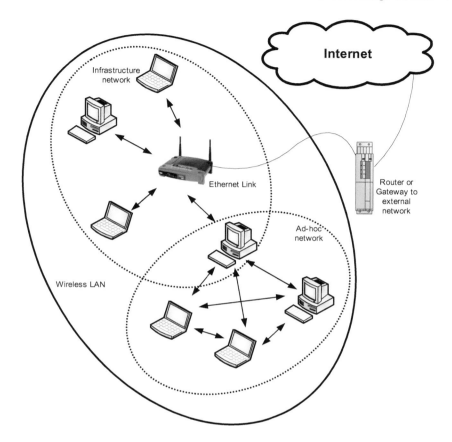

Fig. 1.1 Wireless LAN.

Wireless connectivity technologies having radio range lesser than WLAN is called WPAN. WPAN has range around 10 m. It is generally used in home and office environment for quickly connecting devices with each other and for forming *ad-hoc* networks. It removes connecters and physical connections in gadgets we use everyday to make life simpler. WPAN is used for inter-connecting devices in Consumer Electronics (CE) industry, for connecting mobile phones to other devices like printer, scanner etc. in mobile industry, for connecting peripherals to PC wirelessly in PC industry. Technologies like Bluetooth, IrDA, Zigbee, HomeRF etc. are the players in WPAN. Increasing base of new applications, data rate, streaming, QoS, isochronous, security requirements etc. are driving new wireless technologies in WPAN space. It is targeted for the devices like HDTV, DVD players, Digital Video Recorder (DVR), External hard disk drives etc., the devices with high data transfer requirements and devices with QoS and streaming requirements.

Figure 1.2 illustrates usage of Wireless PAN in different industry segments and its interconnections.

Research is happening in wireless communication, physiological sensing, intelligent monitoring device, light weight, small size, ultra low power integrated circuits (ICs). Wireless Body Area Network (WBAN) is relatively a new technology targeted for connectivity of these devices which can be worn on body and used for health monitoring. Radio range of these devices is around couple of meters. Depending on requirement, WBAN can use connectivity technology of WPAN like Bluetooth, Zigbee etc. for network formation. Other element is a sensing element which could be a set of physiological, kinetic, and environmental sensors. These wearable devices with different kind of sensors continuously monitor different activities of the person and provide the information to the central unit. Figure 1.3 shows different examples of wearable

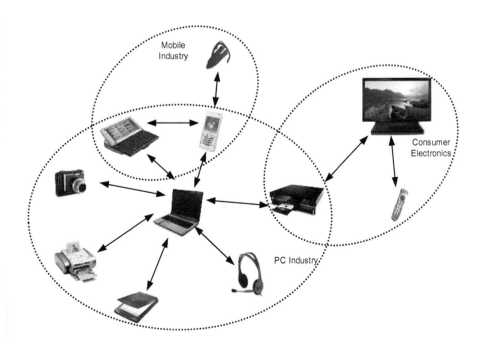

Fig. 1.2 Wireless PAN.

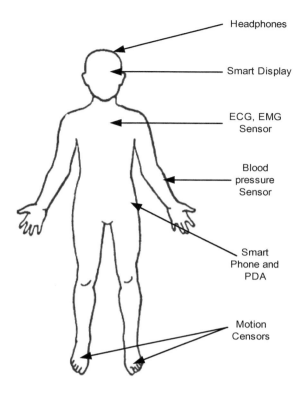

Fig. 1.3 Wireless BAN.

devices and Wireless BAN. So, the WBAN is a telemedical system which can be a key infrastructure for remotely supervised personal rehabilitation.

Wireless Sensor Network (WSN) is a new class of wireless network that is used for getting the properties and state of the physical condition where the sensor device resides. This category is based on usage of the short range wireless and sensor entity. It is used in home security and surveillance, environmental protection, highway monitoring, and military applications. Wireless sensor network is also used in monitoring building, utilities, industrial area, shipyards etc. It is also used for network based identification. Technologies like RFID, Near Field Communication (NFC), Zigbee are the widely used ones. Figure 1.4 shows an example of home sensor network. Different technologies of wireless with various topology formations are used in this for monitoring temperature, sound, vibration etc. It is expected to work in adverse conditions,

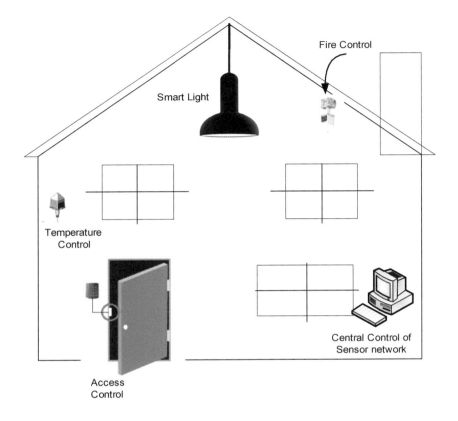

Fig. 1.4 Home sensor network.

so robustness and unbreakable connectivity are few key parameters for sensor networks. Lots of research and standardization efforts are going on for protocols of sensor network and wireless connectivity for the same.

1.3.1 Evolution of Wireless Personal Area Networks

As need of mobility and personal telephone access irrespective of location increased in the mid 20th century, cellular network became wireless extension of wired telephony. Different standardization efforts were done for cellular network in different alliances at international level. Soon after initial cellular network deployment, people felt requirements for shorter range wireless, the wireless for replacement of wired LAN and it led to WLAN. IEEE successfully created WLAN standard as IEEE 802.11. Similarly HiperLAN also created the standard for WLAN in European region to initially use in office environment LAN interconnection.

With progress in wireless technologies, new applications like *ad-hoc* networking, different applications of device interconnection, and applications with localized perspective increased. These applications are generally of very short range and for home networking, unlike office networking through WLAN. New requirements helped to emerge new wireless technology called Wireless Personal Area Networks (WPAN). Initially, there were different proprietary implementations present. As time progressed, interoperability and coexistence started becoming fundamental issue for robust and stable products and technology to be successful. So, for these proprietary technologies it was not easy to reach to broader market. Unlike WLAN, WPAN is a traveling network, a network which can move with person to different locations, airports, different countries, so security and international regulatory is an important factor for WPAN. To tackle many of these issues, solution was sought in the standardization of WPAN.

Bluetooth special interest group (BT SIG) was the first to come with standardized WPAN. IEEE started with new study group for WPAN, because existing standards group for wireless LAN 802.11 was not deemed suitable for requirements of WPAN. WLAN and WPAN are different in terms of radio range and applications. Usage scenario of WPAN is also quite different from WLAN. After initial analysis by study group, IEEE finally started new standards work group IEEE 802.15. Aim for 802.15 was to define PHY and MAC standards for WPAN. Later different types of WPAN PHY and MAC with different characteristics were included in scope of the work group. With time progressed, IEEE adopted Bluetooth under IEEE 802.15.1. IEEE 802.15 also included high speed WPAN PHY, Low rate WPAN, Mesh networking and Body area network in its scope. IEEE, with this initiative, and many other industry alliances helped to define standardized WPAN, which can coexist and interoperate. This helped WPAN reach to broader marketplace. New innovations in WPAN still continue with different data rate requirements, new applications, and different usage scenario.

1.3.2 Bluetooth

Bluetooth is one of the most popular and successful technology for WPAN. It is a connectivity system for low data rate short range communication among fixed or mobile devices in WPAN. Today, we see lots of devices enabled with Bluetooth and very successful for low data rate applications. In fact, Bluetooth has become one of the most sought after features in any cell phone. Bluetooth is used to connect and transfer the data among mobile phones, telephone, laptops, PC, PDA, printers, cameras, mobile hands free, etc. Bluetooth generally has radio range around 10 m and operates over 2.4 GHz frequency band. It is based on frequency hopping spread spectrum technology. One of the most commonly used applications of Bluetooth is voice transfer over mobile hands free kit.

Bluetooth is originally developed and licensed by Bluetooth Special Interest Group (SIG). Bluetooth SIG includes many companies from Telecommunication, PC, Networking, Mobile and Consumer Electronics market segments.

Bluetooth is low power consumption system for short range connectivity. It generally does not require line of sight for communication, if transmission is powerful enough. It provides data rate up to 3 Mbps so it is useful for the applications that does not require high data rate. It is used in applications like low speed networking, connection of mouse and keyboard, connection between PDA and mobile phone etc.

Bluetooth operates as centralized piconet based network where master device, as coordinator, provides many services and manages the network. Master device supports seven active devices at a time connected to master and enjoying the Bluetooth services. More than one piconet can form a network called scatternet, where a bridge device that can be part of both the piconets act as master as well as slave. That device acts as bridge between two piconets. Figure 1.5 shows network topology of Bluetooth called scatternet, which includes two piconets.

Many new generation PCs and laptop computers include Bluetooth master device. For the master device in PC, there are cards and dongles available which can convert wired interface to Bluetooth wireless interface.

Bluetooth SIG defined entire protocol stack for Bluetooth that includes Bluetooth radio, Baseband, Link Manager Protocol (LMP), Logical Link Control and Adaptation Layer (L2CAP), Service Discovery Protocol (SDP). Using this protocol stack, many upper layer protocols are adopted to wireless link of Bluetooth. A low power version called Wibree is being proposed by Nokia and others to be used in cell phone where power consumption is a key target (Refer [12]).

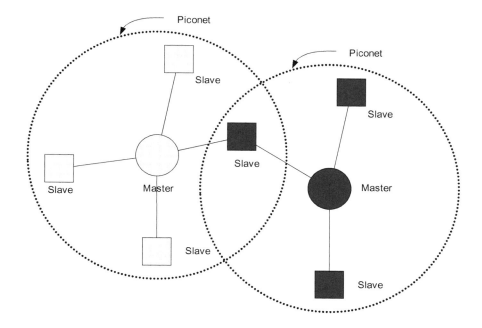

Fig. 1.5 Bluetooth scatternet.

1.3.3   Zigbee

Zigbee is another WPAN technology that is used for data transfer using high level communication protocol with low power. This technology provides comparatively simpler and cheaper WPAN connectivity, with support of low data rate, with long battery life and support of secure wireless mesh networking.

The Zigbee standard, defined by Zigbee Alliance, also publishes the application profiles. Using these standards, implementers can develop the products which can interoperate with each other. IEEE 802.15 being an initiative for WPAN, IEEE 802.15.4 accepted Zigbee for low data rate WPAN.

Zigbee operates in ISM bands; 868 MHz in Europe, 915 MHz in countries such as USA and Australia, and 2.4 GHz in most of other countries. Typical application areas for Zigbee are home entertainment and control like smart lighting and movies, music control; home sensors like water sensor, access sensor etc.; mobile services like mobile payment and mobile security; applications in commercial building like lighting and access control; applications in industrial plants like asset mangement and device controls.

Zigbee is centralized WPAN with a coordinator. It supports three different types of devices. Zigbee coordinator is the manager for the network. Each Zigbee network is started by the coordinator. It has all information about the network, about services, about connected devices. Another class of device is Zigbee router that routes the traffic from one device to another. It can also run the normal applications on itself. Zigbee end device runs the application and communicates to either coordinator or router. It supports functionality like ability to sleep in order to save the power and live long battery life. Zigbee network configuration could be different kind of topologies like star, cluster, mesh network etc. Figure 1.6 shows Zigbee topology.

Due to its low cost, Zigbee can be deployed easily in high number of nodes, so it is very well suitable for many control and monitoring applications. Low power consumption allows it to run for long time on very small batteries. Also, the mesh networking capabilities provides high reliability and larger range for the communication.

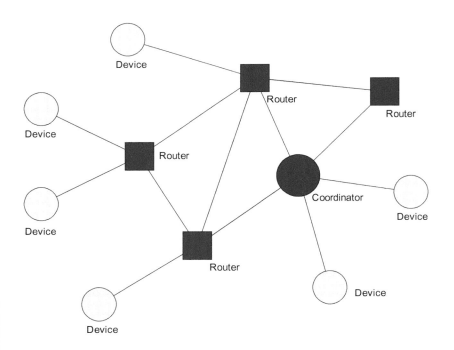

Fig. 1.6 Zigbee topology.

1.4 Organization of the Book

The book mainly focuses on emergence of UWB, evolving standards for UWB, practical system design/implementation and real life applications. It also provides the information about the history of UWB. The book also includes notes on the standards development efforts, regulatory constraints, industry trends, and power management techniques essential to its usage in portable handheld devices.

This book is organized as follows:

Chapter 1: Introduction to Short Range Wireless. This chapter introduces different wireless technologies and their intended applications/usage. It introduces the short range wireless technologies, its use cases and briefly describes some popular wireless technologies like Bluetooth and Zigbee.

Chapter 2: Ultra Wideband. UWB was earlier used in military applications. With FCC's approval to commercial use, it created opportunities for new wireless applications. This chapter describes UWB signal characteristics and possible use cases of new age UWB connectivity.

Chapter 3: Evolution of UWB Standards. Since UWB is approved by FCC for commercial use, different companies and standard bodies attempted to create standards to use it in commercial application. Standardization is a key for any successful technology to ensure interoperable and robust products. This chapter describes the efforts of IEEE, other industry alliances and organizations for UWB standardization.

Chapter 4: Physical Layer. This chapter describes the Multi-band orthogonal frequency division multiplexing (MB-OFDM) based Physical layer for UWB system adopted by ECMA-368 standard, one of the most popular and attractive in terms of high bit rate, low power, short range, and cost-performance ratio.

Chapter 5: Medium Access Layer. UWB system requirements and application base are quite different from other existing wireless. Medium Access Layer design is influenced by the requirements of UWB system. This chapter describes various design choices/constraints for MAC, details of ECMA-368 MAC control algorithms and data transfer mechanisms for UWB system, which is suitable for several upper layer protocols over wireless link.

Chapter 6: *Advanced MAC Features*. UWB being wireless connectivity technology for handheld and mobile device, power save and security are very important features. This chapter describes power save and security features provided ECMA-368 MAC (UWB MAC) with unique features of ranging for UWB Systems.

Chapter 7: *UWB System Design*. UWB system is a unique system with specific requirements. Various kinds of system models are possible to achieve these requirements. Multiple upper layer protocols can be active simultaneously over the system. This chapter addresses these requirements and suggests how to achieve those.

Chapter 8: *Adaptation to Multiple Applications*. UWB common radio based on ECMA-368 is extendable system which can be used for many different upper layer protocols. MAC provides various tools like frames and multiplexing mechanism for application specific usage. This chapter provides information about how UWB system can be used for different applications.

Chapter 9: *Wireless USB*. Wireless USB unwires popular and traditional USB connectivity over UWB. WUSB is protocol adaptation layer over UWB common radio platform. This chapter provides details about how UWB common radio can be used for WUSB and details of WUSB protocol.

Chapter 10: *Converging Marketplace.* This chapter covers the marketplace of UWB devices, discusses new applications of UWB, and other competitive technologies like 802.11n. It also discusses the Test and Measurement instruments for UWB. Discussion also covers the need and applicability of various wireless technologies when "all in one" devices are going to be an increasing trend.

2

Introduction to Ultra Wideband

2.1 Introduction

Data transmission technology over wide frequency band and employing large bandwidths is called Ultra Wideband (UWB). Ultra Wideband is merely a frequency band. Technologies that are proposed to use this band are referred in literature as UWB technologies. Using this radio frequency and sharing the spectrum with multiple users, this technology provides support for high data rate communication for short radio range. This makes it suitable for high speed WPAN.

Ultra Wideband is not a new technology. It has been known since years, initially for its usage in military, now for its commercial usage and its initial products. Today UWB is a popular evolving radio technology, which can be used at very low power level over Ultra Wideband radio spectrum. There are many mechanisms for wirelessly transporting voice, video, and data at very high rate over UWB.

In early 2002, Federal Communications Commission (FCC) in United States approved UWB for commercial use, with limits on maximum radiated power levels to ensure that its operation in unlicensed band does not cause severe interference to other players in the given bands. It enabled many high throughput consumer electronics, mobile and PC applications to go wireless over UWB. Regulations of FCC require UWB radio not to interfere with existing users and applications of the same spectrum. Very high bandwidth enables both high data rate Wireless Personal Area Network (WPAN) at short range, and at longer range, low data rate. Traditionally UWB is a pulse based transmission technology. Due to FCC regulations, which limits emitted signal bandwidth that led to nonpulse-based transmission techniques and products based on that. Details about UWB technology, signal characterization, its commercial usage along with history of UWB are explained in the following subsections of this chapter.

2.2   History

As a concept, UWB is quite old with rich history. Hundreds of years back, the groundbreakers of wireless communications and electro-magnetic wave generation (Heinrich Hertz, Stepanowitsch Popow, Guglielmo Marconi) have used UWB signals. Guglielmo used large bandwidth sparks when he transmitted and received first electro-magnetic signals during the year 1894.

With more and more people using radio transmission and multiplicity of radio technologies, the remaining free spectrum had got scarce, and allocation for frequency bands started becoming constrained. Researchers worldwide got engaged in developing technologies that were spectrum efficient and possibly capable of operating along with other technologies in the same bands. Different countries across the globe including USA, Russia, China have started contributing in area of UWB systems and started building various UWB theoretical models and prototypes. Before commercial development in UWB area kicked off, crucial concepts for signal theory of UWB have been developed which includes pulse compression, correlation, match filtering etc. and were already in place for ready to be used in UWB systems.

In earlier years, the UWB signals were used as an umbrella term for all terms including: baseband, time domain, impulse, carrier-free, nonsinusoidal, orthogonal functions and large relative bandwidth radio/radar signals. Development in the area of UWB systems initiated in around early 1960s. With the pioneering work done by Harmuth at Catholic University of America, Ross and Robbins at Sperry Rand Corporation, Paul Van Etten of USAF's Rome Air Development Center. Ross' US Patent 3,728,632 (Dated 17th April, 1973) is considered as turning point in UWB communications. Though the grass roots are in 1960s, the word "Ultra Wideband" is coined recently during 1989 by US Department of Defense.

The Harmuth's books and published papers during 1969 to 1984 have put basic design of UWB transmitters and receivers in public domain. By the 1970s basic design was almost available. However, adaptability and realizations of these has been a key challenge. During same time, the Ross and Robbins have shown the use of UWB signals in number of practical applications, in the area of communications and radar signals. They also demonstrated coding schemes.

Ross and Robert prepared a seminal paper entitled "Comments on Baseband or Carrier-free Communications." Ross and Robert continued collaboration on UWB system research and development for both communications and radar applications for approximately 11 years.

Few other innovations accelerated UWB growth tremendously. In 1974, Morey designed a UWB radar system for penetrating the ground, which was commercialized by Geographical Survey Systems, Inc (GSSI). During 1977, Van Etten's empirical testing of UWB radar systems has given thrust to the development of system design and antenna concepts.

Technology leaders Hewlett-Packard and Tektronix have accelerated UWB implementation by making oscilloscopes as well as fast sample and hold receivers commercially.

In last decade, when UWB is considered for commercial usage, regulatory forums started defining regulation and constraints for UWB signal usage, many different modulation schemes have been developed and product prototypes being realized in a journey to see UWB based products in our hands.

2.3 Signal Characterization

As name suggests UWB is wide bandwidth technology, significantly different from traditional narrow band radio frequency technology and spread spectrum technology.

Traditional wireless systems like 802.11a/b/g and Bluetooth are narrow band technologies. Compared to narrow band technologies, UWB can transmit more data over wideband. Data rate over radio frequency depends on bandwidth and logarithm of Signal to Noise Ratio (SNR). Bandwidth restrictions are imposed by regulatory of different countries. Different technologies like Bluetooth, 802.11, and others have frequency band provided are 900 MHz, 2.4 GHz, or 5.1 GHz with narrow band of approximately 80 MHz around the center frequency of a radio.

Comparatively UWB is provided really wideband from 3.1 GHz to 10.6 GHz, where each radio can have 25% bandwidth around center frequency. So, minimum bandwidth of 500 MHz and more than that is used in frequency spectrum.

The most common technique for generating a UWB signal is to transmit pulses with durations less than 1 nanosecond. Because of short duration of the pulses, the frequency spectrum of a UWB signal can be many GHz wide, overlaying the bands used by existing narrow band systems. The interval between pulses can be uniform or variable. Numbers of different methods are defined for modulating the pulse train with data for communications over UWB. It does not require carrier frequency so it is called zero carrier radio also.

Narrow band transmitter broadcasts considerably high power for reliable reception at receiver side. With this power level it can go up to the radio range of around 100 m. Range and power levels are strictly defined by regulatory. Figure 2.1 shows UWB spectrum with the other spectrums that used narrow band technologies with their power levels.

As a result of UWB's distribution of energy, the spectral density is extremely low. UWB providing such a high bandwidth, regulatory has put restrictions on power broadcast. Because of these restrictions UWB transmissions are not detected by nearby narrow band receivers, but these power restrictions limit the range of UWB radio to few meters, which is suitable for WPAN only. The constraint imposed by FCC is that energy spectral density does not exceed more than -41 dBm/MHz. Signal spectral density drops sharply to -60 dBm/MHz at frequencies below 3.1 GHz. The frequency band of less than 3.1 GHz is not used by UWB because of other narrow band technologies

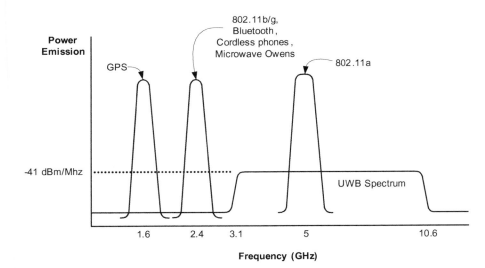

Fig. 2.1 UWB spectrum.

using that band and it can create interference to any other signal in same band though at very low signal level. That is why 20 dBm, from −40 to −60 is reserved to not create interference to narrow band signals and also used in military applications.

Besides UWB applications in radar, UWB can send lots of data when many simultaneously acting radios are at shorter distance to each other. Density of transferred data able to coexist on the same square meter is much higher for the UWB compared to the narrow band systems. Requirement of controlled transmission of power and limitation in radio range improves the spatial effectiveness of the spectrum utilization in commercial usage of UWB. Traditional narrow band modulation technologies require guard bands for eliminating pickups that leads to low spectral utilization.

UWB radio, employing shorter pulse durations, has ability to detect the distance to another UWB radio based on round-trip delay information more accurately. Range information can be used by device to further reduce the transmit power and can optimize the network capacity and traffic flow.

Being shorter pulse durations, UWB is also highly immune to the effects of multi-path. Signal bandwidth of a UWB signal is similar to the coherence bandwidth of the multi-path channel. So, there are many possible paths that a UWB radio can use for communication that makes UWB very robust.

Ultra Wideband has many advantages which makes it suitable for indoor short range wireless with very high data rate for consumer-type applications. Due to physics of UWB, it has an inherent ability to go through walls and in cluttered high multi-path environments.

So, UWB provides high throughput at short range for numerous WPAN applications, which are explained later in this chapter.

2.4 Ultra Wideband in Wireless Personal Area Networks

Wireless Personal Area Networks were introduced earlier in this chapter along with brief discussion on different existing WPAN technologies. Next, we discuss the requirements of WPAN resulting from next generation applications.

Low power consumption, high throughput with QoS support, quick *ad-hoc* networking facility, connectivity with infrastructure network, large number of devices support, support of various type of traffics, operation without line of sight etc. are the essentials for WPAN. Also not forgetting the basics of any wireless mobile technology to be popular are low cost, low complexity, and small size.

Ultra Wideband having short range can provide very high throughput of hundreds of Mbps. It comes with low cost and power with almost digital architecture. Low signal energy minimizes the interference to other devices. It is extremely difficult to intercept and is immune to multi-path cancellation. It is able to pass through solid objects like building blocks and wells.

All these characteristics make UWB suitable for high speed WPANs. Other than being wireless replacement of wired connection, UWB can provide services to many different segments of industry.

2.5   Use Cases for Ultra Wideband

Ultra Wideband is suitable for many of commercial products like PCs, printers, scanners, digital cameras, media players, HDTV, storage devices etc. Broadly considering market segments, UWB can be used for:

(a) PC industry, which includes PC wireless connectivity with PC peripherals like mouse, keyboard, and devices connecting to PC like printer and scanner.

(b) Consumer Electronics industry, which includes HDTV, media player, speakers, digital cameras etc.

(c) Mobile phone industry, which includes WPAN connectivity among mobile phones, headphones, PDAs etc.

In current era of heterogeneous network and high multimedia applications, all these market segments are not really separate from each other. These market segments are connected with each other. UWB has great opportunity to serve as connectivity medium among these devices and market segments. Figure 2.2 shows the converged marketplace and connectivity through UWB.

PC Industry: PC industry includes the PC, its peripherals and devices getting connected to PC. UWB can be used to connect peripherals wirelessly like mouse, keyboard etc. It can be used to connect other devices like printer, scanner, digital camera etc. with PC without any wires. UWB is a great wireless replacement of wire in PC industry. UWB can also be used for creating wireless network among multiple PCs.

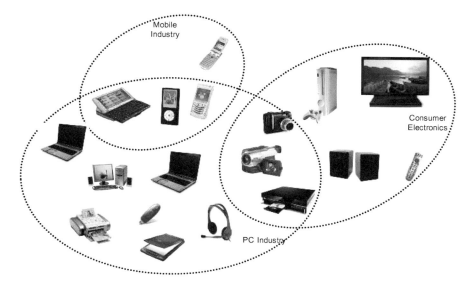

Fig. 2.2 UWB marketplace.

Consumer Electronics Industry: In current era all industry segments are not separate, there are lots of overlap among PC, consumer electronics and mobile industry. Consumer electronics segment includes devices like digital TV, HDTV, video player, video recorder, digital camera, camcorder etc. UWB link can be used to transfer data among these devices.

Mobile Industry: Mobile industry includes mobile and handheld devices, where wireless connectivity with mobility is the key aspect. UWB can be used for short range connectivity between wireless devices like mobile phone, PDAs, tablet PCs, etc. These devices can connect to other devices like PC, printer, scanner, TV etc.

2.5.1 PC and Peripheral Applications

PC has always been center to many peripheral or gadget connectivity and host of many applications. Numerous peripherals, consumer electronics devices, external drives, and smart gadgets can connect to PC (e.g. keyboard, mouse, webcam, printers, scanners, digital camera etc.). Even those can become part of network through Ethernet and dialup modems. PC also has traditional wireless

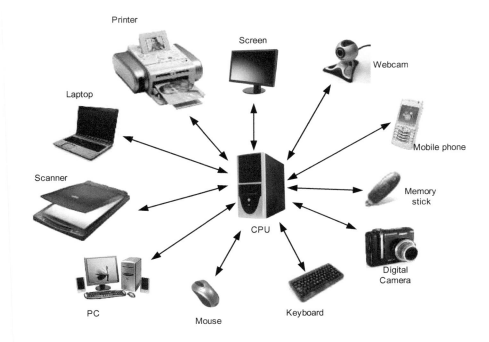

Fig. 2.3 PC and peripheral connectivity through UWB.

connections like WiFi and Bluetooth. Figure 2.3 shows UWB connectivity between PC and its peripherals, as well as other devices connected to PC.

In today's generation, different applications running on PC which has requirement of high speed wireless connection are large. Files on PC hard disk drives can be shared among devices, PCs or laptops on UWB wireless network. Pictures can be downloaded from digital camera, scanner, or mobile phone to the host PC; documents can be printed on printer; keyboard and mouse can be connected wirelessly; video on PC can be played on TV; external memory drives can be connected wirelessly to PC and data can be read and written on that; and many other high speed connectivity achieved through UWB.

Many printers today also have capabilities to print directly from digital camera, mobile phones, or PDAs, so UWB enabled devices can interconnect without PC and perform many ubiquitous tasks very quickly.

2.5.2 Mass Storage Applications

Digital picture and video qualities are improving everyday, but at the same time size of those files is also increasing. People started maintaining lots of data for different purpose. One is always running out of memory on hard disk drive on PC or laptop. Also it is always required to move the data from one computer to another so, external mass storage has become very important in today's systems. Storage is never enough for any PC or laptop, because applications using storage keep on increasing. External pen drives, hard disk drives etc. can be connected to PC. UWB is best suited to replace these wired connections to wireless.

Other devices like digital camera, camcorder, printer, scanner, mobile phone etc. also include mass storage for storing different information like picture, video etc. These devices are also connected to PC or any other devices to transfer the files for printing or viewing. These transfers can be done wirelessly over UWB. Figure 2.4 shows how UWB can connect PC to different external mass storage devices and also other mass storage devices can connect each other.

2.5.3 Streaming Applications

Ultra Wideband connectivity provides very high data rate, so streaming is possible using wireless link of UWB. By suitable MAC design, it is possible for UWB to support for isochronous data transfer for the applications like high definition TV. Physical layer data rate of 480 Mbps can give application layer throughput which can transfer multiple HDTV stream over the air isochronously. So, UWB can be used for playing high definition video over HDTV from media player along with audio transmission to speaker, creating the true theatre environment without any mess of wires. UWB can be used in game consoles where very quick real time data transmission is required.

Ultra Wideband can also be connectivity between set-top box and Digital Video Recorder (DVR), when live stream from set-top box wirelessly goes to DVR and it is recorded real time by DVR. Figure 2.5 illustrates the usage of UWB for interconnecting different streaming devices (stream transmitter and receiver).

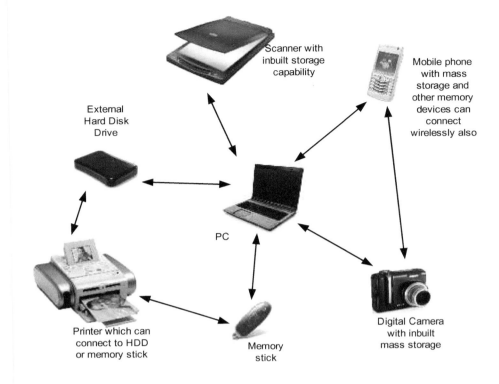

Fig. 2.4 Usage of UWB in mass storage.

2.5.4 Infrastructure Networking

Since long WiFi was popular for wireless networking with support of connecting to backbone wired network or any other wide area network. This enables wireless internet usage and reach to the wide area network. UWB being WPAN provides short range wireless connectivity, but it can also be connected to backbone network. Routing and bridging functionalities can be adapted to UWB that can provide wide area network services access to WPAN network devices wirelessly. This enables internet access wirelessly through UWB.

Figure 2.6 shows the UWB network topology which has connectivity to outside world through router and internet connectivity can be enabled through that.

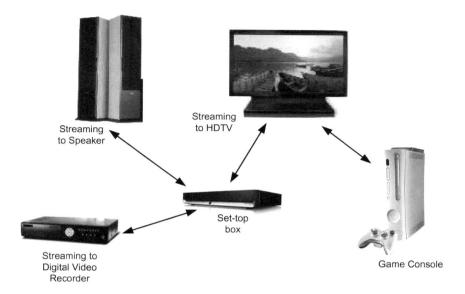

Fig. 2.5 Usage of UWB for streaming.

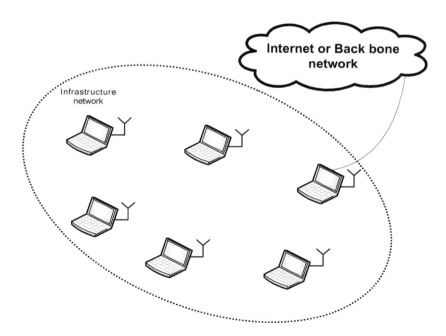

Fig. 2.6 UWB in infrastructure network.

2.5.5 *Ad-hoc* Networking and Mobility

In mobile devices scenario, it is very important to keep the connection alive while device is mobile. UWB is high speed connectivity solution which supports mobility and rapid change in topology. It keeps devices connectivity with each other as long as those are in radio range of each other.

Further in mobile environments, many devices come and go at a time, many devices come in range, go away and come again in range. Those devices need to be quickly synchronized and remain synchronized. For communication between those devices, *ad-hoc* network can be created quickly, many devices can be part of *ad-hoc* network, and many devices may leave the network soon. UWB WPAN can provide these services with data communication among the *ad-hoc* network devices. Quick connection and synchronization are therefore essential elements of the UWB WPAN.

Figure 2.7 shows an example of *ad-hoc* network inside a room, where many devices present in *ad-hoc* network, also a device is coming inside the room which can be part of *ad-hoc* network, another device is leaving room and leaving *ad-hoc* network. With these changes also network remains and remaining devices communicates with each other.

2.5.6 Future Applications

First target for UWB is to provide support of many existing applications wirelessly being wireless replacement of wired connectivity and network, also it is expected to support newly discovered applications.

Applications throughput requirements is also increasing every day, new applications are continuously discovered and physical data rates continue to increase. UWB systems are required to meet those throughput requirements with improvements in the system, so it can support new kind of applications with high throughput requirements.

Not considering only connectivity and high throughput large data transfer, UWB can be explored in other directions like sensor networking and ranging kind of unique applications.

High streaming requirements of HDTV and future generations of HDTV have to be supported by UWB with improved performance in streaming and data rate. Real time system requirement for wireless game stations and quick

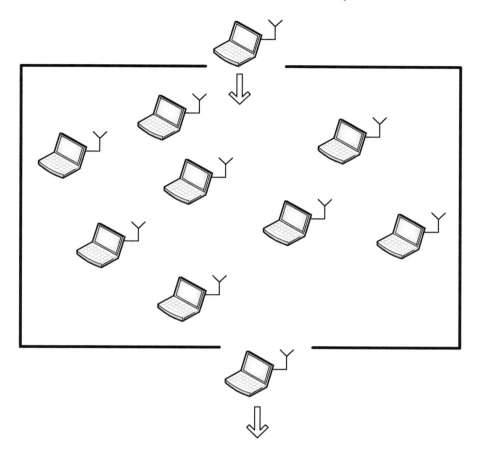

Fig. 2.7 *Ad-hoc* network based on UWB.

response requirements with reliability can give an entry to UWB for large automotive marketplace.

There are companies demonstrating high data rate up to 1 Gbps proprietary implementations and applications using these. Extending Moore's law in data rates, those are expected to be default rates in future and may go higher than that.

Last but not the least; the developers must improve the standards so it can support higher data rates and new application support. Designers, implementers must have to keep the cost of the ICs and systems minimum, with highest throughput and lowest power consumption, so UWB can remain ahead in marketplace and can be adopted by many different markets with new application base.

2.6 Striving for a Standard

Since UWB was approved by FCC for commercial usage, many companies and organizations started working on different methods for data communication over UWB. Different physical layer and MAC designs derived from UWB research and prototypes were developed by different companies. Initial products in market using UWB were available with high data rate but different communication technologies and these products did not work with each other and interoperate because different methods were used for physical layer and MAC layer implementations. Products based on different communication technology near to each other made mess on air and none of those could work at all due to interference.

As time passed, different industry alliances formed and they come up with standards for UWB communication in their own way ostensibly to serve the interest of their members/partners. Dream of transferring huge multimedia data with very high data rate at very low power was impossible without single standard. Interoperability issues due to multiple parallel standards can create situation where none of the products work in presence of each other. Also multiple companies and unions work for their technology and progress of whole market would be really slow because of decentralized efforts. Without standard and authorized certification body, qualities of the products are difficult to judge and maintain.

Eventually, the marketplace without standard can be a threat of killing whole UWB technology. During this turmoil, IEEE started work group for High Speed Personal Area Network Communication (IEEE 802.15.3), where UWB can be considered as high data rate, short range wireless system.

2.7 Summary

Ultra Wideband being a very old technology in military domain is now being considered for commercial usage. How UWB evolved with time and different domain with help of many individual, governments and alliances are explained earlier in this chapter. UWB is a wideband signal which can be used for high data transfer at short range and very low power. These advantages enable different kind of applications. This chapter described usage of UWB in different industry segments like PC, consumer electronics and mobile industry along with different immediate applications and future applications. Standards are required for any wireless technology to be successful. Standards efforts of using UWB for commercial usage are going on, which is explained in detail in the following chapter.

3

Evolution of UWB Standards

3.1 Introduction

Different standards exist for different kind of technologies and processes. Standards are defined to provide uniform technical methods, processes, and understanding. Standards enable multi-vendor interoperability and allow competitive products in market quickly from many different vendors. Especially in wireless communication systems, standards provide interoperability and interfaces for components and products from different vendors, so products can coexist and co-work with each other without any problem. There are different types of the standards and standardization bodies present, like IEEE, ISO, ANSI, ECMA etc. There are many industry alliances which provide their own standards. Some standards are *de facto* standards, those are not defined but are being used since long or driven by market.

One might have a question, why are the standards required? Answer is long and has many reasons. Standards enable vendors to come out with the product in market really soon, which can interoperate and interconnect with other vendors product. Before that it provides opportunity to companies to understand the requirement of different end users from the technology, and to understand the market in broader sense. Standards reduce the risk of multiple implementation of similar technology which cannot interoperate. Multiple implementation scenarios can cause serious impact on success of the technology.

Products can exist in market without standards also. Those are not certified by any standardization or certification bodies. Suppose one brings home such product from market and soon finds that it works well with same vendor's product. But, with another vendor's product it does not work well or may not work at all. This explains why standards are required in end user term. In summary, standards are required to meet the market requirements in best way, on time, with best possible methods or processes.

Since after Federal Communications Commission (FCC) approved Ultra Wideband (UWB) for commercial use many companies started working on prototypes and commercial product using UWB. It includes major semiconductor companies and small startup companies. Same time IEEE formed a study group under IEEE 802.15 for high speed short range wireless called SG3a, UWB was considered for this group. Expectations from this group was to develop standard for WPAN which can provide sufficient quality of service for the real time distribution of content such as video and music, which could be suitable for home wireless multimedia. Under the umbrella of 802.15 there are multiple WPAN technologies covered, like task group 1 is Bluetooth, task group 2 addresses the issue of coexistence, task group 3 is high data rate WPAN (including MAC and PHY), task group 4 is low data rate WPAN (Zigbee used standard from this task group for basic connectivity), task group 5 is mesh networking group, task group 6 is for body area network.

During the initial time of the technology many companies had many ways for implementation of UWB for WPAN. There were companies who demonstrated their implementations and different high data rates. Some were most concerned about the new usage models and use cases of UWB, what all will be the end applications for UWB, what will be the targeted devices. While some people wanted standards to be written in their way so they can go for productions and hit to the market as soon as possible, there were groups that wanted to do feasibility study, market research, so technology can be used for right applications. Looking at the pace of the technology development, there was a risk of the multiple standards, which can fragment the marketplace and could make the technology a failure.

3.2   IEEE Standards Dilemma

Original IEEE 802.15.3 standard was using traditional carrier based 2.4 GHz radio as physical medium and piconet based centralized MAC for WPAN channel management. To meet the requirement of using 802.15.3 system for wireless home networking with low complexity, low cost and low power consumption, group had attempted to provide higher speed UWB PHY through IEEE 802.15.3a task group. Same time, 802.15.3b started working on an amendment of 802.15.3 MAC for better interoperability and implementation over new proposed high speed PHY and clarify some errors and ambiguities.

TG 802.15.3a worked on UWB PHY standard for more than three years. Group reviewed almost 23 different UWB PHY proposals and achieved two final proposals after mergers and consensus; both were supported by two different leading industry groups. One proposal was Multi-band Orthogonal Frequency Division Multiplexing (MB-OFDM) UWB, supported by Multi-band OFDM Alliance (MBOA), which merged with WiMedia Alliance later and called WiMedia Alliance. Another proposal was Direct Sequence-UWB (DS-UWB), supported by UWB Forum.

WiMedia Alliance is an open industry association that promotes rapid adoption, regularization and standardization of UWB for high speed wireless for multimedia capable PAN in consumer electronics, Mobile and PC market. WiMedia UWB PHY was optimized for wireless personal area networks delivering high speed (480 Mbps), low power multimedia capabilities for the PC, CE, mobile, and automotive market segments. WiMedia was led by different PC and CE companies like Intel, Staccato, Wisair, WiQuest, Samsung, Sony, NEC, Texas Instruments, and Philips.

The UWB Forum, led by Freescale, was an industry organization comprised of leading semiconductor, software, OEM/ODM, and consumer product companies dedicated to ensuring that UWB products from multiple vendors are truly interoperable. DS-UWB PHY also provided high speed data transfer, which can be used for mobile phones, set-top boxes, computers, televisions, digital camcorders, and more.

After nearly three years of debates on technology and process issues, UWB PHY standardization attempt failed in IEEE due to contrast between proposals supported by WiMedia Alliance and UWB Forum. Each of the remaining proposals contained mutually exclusive communication architectures. Also after few years of improvements by both the group, both proposals achieved nearly identical performance in terms of data throughput, channel robustness, power consumption, and device cost. Due to different basic modulation techniques of PHY neither proposal's radio could communicate with the others, so both the radios cannot coexist. There were numerous attempts by both the groups to achieve victory in the down selection vote in IEEE, but it was never achieved. There was an attempt for compromise by creating a Common Signaling Mode (CSM) channel using lower data rate for coordination, coexistence, and minimal interoperability. But this was also a failure, because according to market research of MB-OFDM supporters, only one PHY can be tolerated by the consumer market. Finally, 802.15.3a task group decided to close the work on UWB standard and decided to leave the issue of PHY to market.

After coming out of IEEE, each side remains committed to its technology and continues pressing forward. The UWB Forum was marketing a DS-UWB technology that was incompatible with the MB-OFDM technology backed by the WiMedia Alliance. While performances may be comparable, products made with the two technologies may not be compatible or interoperable. It is possible that only one of these technologies will survive in next few years and IEEE can come back again and think of creating a standard from surviving technology.

In year 2006, the leaders of UWB Forum, Freescale and Motorola decided to quit UWB Forum to focus their efforts on product development based on UWB, rather than standards development. They thought that the forum was de-focused with concerns such as regulations, certification, interoperability, and common signaling.

3.3   MBOA and WiMedia

While IEEE was in deadlock for UWB PHY, MBOA had finished their draft specification for UWB PHY which was also a proposal in IEEE 802.15.3a. MBOA also was in process of defining new MAC layer over MB-OFDM PHY which can suit to various kind of WPAN application and in turn, this happened to be a radically different approach from IEEE 802.15.3b MAC. Same time WiMedia Alliance was defining a MAC convergence layer specification over UWB radio, so it can be used by different upper layer like Wireless USB, Wireless 1394, IPv4/v6. WiMedia was also defining protocol adaptation layer for IPv4 and IPv6, was called WiNET, eventually it was called WiMedia Link Protocol (WLP). At that time WiMedia was acting as a neutral party for UWB similar to the Wi-Fi Alliance for IEEE 802.11.

Eventually, WiMedia and MBOA decided to merge in early 2005 and provide a whole UWB system specification set, which can cater various WPAN applications at very high speed. Also there was tremendous overlap of membership between WiMedia and MBOA; merge was a logical business move beneficial to both the parties. These reasons made merger easy and natural. Both the groups (MBOA and WiMedia) were working on different standards development and those were continued as part of WiMedia and finally they delivered, PHY, MAC, and WLP specification. PHY was defined by MBOA which was taken in by WiMedia. MAC was started in MBOA and continued in WiMedia. WiMedia MAC convergence layer specification eventually merged with MAC specification. WiMedia defined Protocol Adaptation Layer (PAL) for IPv4 and IPv6 in form of WLP.

WiMedia has liaison with Universal Serial Bus Implementers Forum (USB-IF) and Bluetooth SIG, respectively, where they defined WUSB and PAL for high speed Bluetooth.

Merged group also started defining its certification and interoperability policy including certification and interoperability specification for PHY, MAC, and WLP. At this time, WiMedia has certification and interoperability process and tools in place to certify WiMedia PHY and Platform with MAC and PHY. WiMedia also took the MAC–PHY interface specification defined by MBOA for standard interface for PHY chip vendors. Also, merged group continued the battle in IEEE for MB-OFDM proposal.

This merger, one more consolidation in UWB space could not solve the deadlock in IEEE. Not waiting for solving deadlock in IEEE, WiMedia continued its work toward UWB system and provided the specifications and that architecture for UWB in what is commonly called UWB common radio platform, which can be used for different upper layer through protocol adaptation layer, either provided by WiMedia or its liaisons.

Toward the end of 2005, European Computer Manufacturers Association (ECMA), called ECMA International has released UWB standards, ECMA-368 (High Rate Ultra Wideband PHY and MAC Standard) and ECMA-369 (MAC-PHY Interface for ECMA-368) based on WiMedia MAC, PHY, and MAC–PHY interface. The ECMA-368 and ECMA-369 standards were approved as ISO/IEC standards in 2007 as ISO/IEC 26907:2007 (High Rate Ultra Wideband PHY and MAC Standard) and ISO/IEC 26908:2007 (MAC–PHY Interface for ISO/IEC 26907). Now, ECMA-368 is also an ETSI standar as ETSI TS 102 455. Because of adoption of WiMedia UWB by different standards bodies, it got international industry acceptance leading the path toward some product development for the emerging marketplace.

3.4 Regulatory Considerations

Due to the nature of wireless, radio systems can interfere to each other when multiple systems are near to each other. These interferences are potential enough to break the system, applications, and security also. Even it can lead to industry concern for multiple wireless technologies. Technology must be designed and developed for flexible usage across multiple countries and associated regulations.

When FCC defined UWB, it was defined only as a spectrum that was made available. It also defined the power spectrum profile describing maximum radiated power (peak power emission limits) for each frequency within the permissible bands to ensure limiting interference to other devices within this "unlicensed" operation. In addition, FCC also imposed separate frameworks for UWB usage in both indoor/outdoor applications and restricting UWB usage in specific scenarios. For example, UWB operation is not allowed on the aircrafts/satellites to limit any unwarranted interference.

As the UWB standardization and development has progressed, several other countries have also been in the process of defining spectrum bands and other necessary regulations for UWB deployment. This has been, at times, in conjunction with other emerging technologies such as WiMAX.

Importance of regulation lies in specifying what a UWB radio, as part of any product, can emit in the given frequency bands. Together with attempts by different countries to specify the frequency bands for UWB usage, the standards have also been evolving to take maximum usage of UWB deployment through flexible and optional components in the specification.

For the purposes of this book, it is not intended to detail various test infrastructures and measurement techniques for the UWB radio behaviors, neither do we intend to focus on various instruments that will be needed to ensure regulatory compliance. Readers are referred to [93] for a more comprehensive treatment on regulations. Instead, we seek to provide informatory notes on the status of UWB spectrum allocation across different countries and the attempts made by UWB standardization work to introduce flexibility in radio operation through emergence of Detect and Avoid (refered as DAA) techniques.

One of the key challenges in deploying UWB as commercial solutions was overcome when FCC allowed a waiver in 2005 toward certain technical requirements. In simplistic terms, this waiver relaxed the requirements on measurements done in case of frequency hopping modes, allowing measurements to be made over longer intervals allowing averaging effect for the transmit radiation levels. The waiver was granted specifically for indoor and handheld

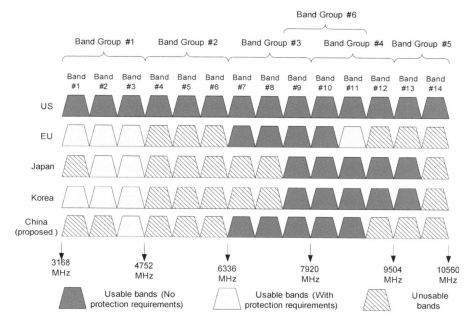

Fig. 3.1 Regulatory spectrum.

devices, therefore paving the way for accelerated UWB deployments in indoor and handhelds businesses.

Figure 3.1 depicts the regulatory spectrum available (including proposed to be available) in different countries. For United States, the entire spectrum in all 14 bands was made available, while European Union and Korea approved only specific bands as shown with specific provisions for incorporating DAA mechanisms. In Korea, the use of unlicensed low power emission devices is restricted such that the upper limit of the allowable emission level is lower than the current FCC regulation by 25 dB. Therefore, without introducing a new regulatory plan, only limited applications of low power devices can be allowed.

For Japan, regulatory approval process has reached the stage of compliance testing being complete and availability of DAA mechanisms in the 4.2–4.8 GHz bands. Band group 6 was specifically proposed to promote universal applicability. The regulatory process for China, Canada, Singapore, and Brazil are underway, expected to be approved by the end of year 2008.

The challenge is to design UWB systems that can cater to different regulatory scenarios with inherent flexibility in the system rather than needing specific devices for specific countries. Because there are more than one band available in each region with at least one common in every region, product manufacturers has the required flexibility in creating products with geographical reach.

There are some bands which are opened for devices employing DAA technology. DAA is one of the techniques for solving the interference due to UWB transmissions. When UWB device detects a signal from a non-UWB device, it can avoid causing interference by reducing its signal strength. Details of the DAA are described in Chapter 4.

3.5 Common Radio Platform

WiMedia UWB Common Radio Platform incorporates MAC and PHY layer based on MB-OFDM. Figure 3.2 shows basic architecture of UWB common radio platform and upper layers of that.

ECMA-368 based UWB network architecture is distributed and decentralized network architecture where each device is self contained and each device takes care of itself by learning the neighborhood. This architecture has several advantages, like tolerant of faults in all of the devices and eliminating any possible bottlenecks a centralized architecture can bring. It provides the mobility in terms of change of topology. Self contained UWB device can learn the change in its neighborhood.

Common radio platform is designed in such way that the entire decision making is done through device itself. In case if multiple upper layers are present then also decision making is done by the device considering requirement of different upper layers. Distributed architecture supports multiple upper layer protocol through PALs. It is possible that upper layer or PAL has centralized architecture over UWB common radio. ECMA-368 MAC supports reservation based and as well as contention based data transfers at different data rates with various power save mechanisms. It provides different acknowledgment policies for reliability. It has a unique feature of fairness of the usage of the medium through MAC policies in all situations.

The architecture of radio platform looks complex but provides robustness, flexibility, and fairness. Multiplexing sublayer commonly called MUX

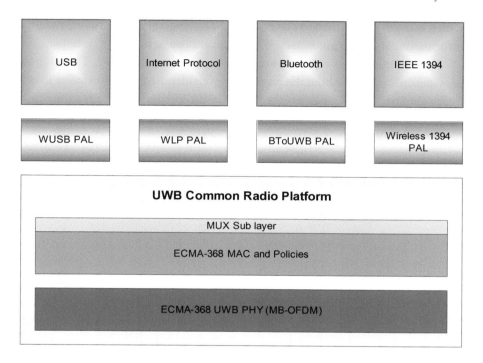

Fig. 3.2 UWB common radio platform.

sublayer of MAC provides multiplexing of the data for different PALs. So, the architecture of common radio platform serves many upper layer protocols and it is extendable to serve other upper layer protocol in future. As shown in Figure 3.2 there are different upper layers like Wireless USB, Wireless 1394, Internet Protocol, Bluetooth, and etc. can be served by same UWB radio.

3.6 Summary

This chapter describes the UWB standards and efforts by different standardization groups and industry alliances for taking UWB ahead. The deadlock created in IEEE between two leading proposals of UWB PHY could not be resolved with time. As a result the two alliances continued their respective work on products.

Common radio platform based on MB-OFDM UWB provides opportunity for different upper layer protocol to use high speed connectivity system, which is briefly explained in this chapter. Following chapters provide detailed information about common radio platform and usage in UWB systems.

Regulatory always effects adoption of any new technology. This chapter provides information about the regulatory consideration and constrains in different countries, which are expected to be initial adopters for UWB technology and products.

4

Physical Layer

4.1 Introduction

The Physical (PHY) Layer deals with processing of signal before transmitting it over the air. The ability to use the frequency band in a spectrally efficient manner is one of the key requirements of the PHY. Evolution of advanced modulation and error control coding techniques has resulted in plethora of choices for the system designers.

In this chapter, we discuss the requirements of UWB PHY layer, requirements from PHY for UWB based WPAN environment, various design choices, transmitter and receiver operations. Since UWB operates in unlicensed band, coexistence with other technologies is one of the important considerations for the PHY layer, which is also discussed here.

4.2 Requirements

The requirements for the PHY layer of UWB was specified by IEEE 802.15.3a Project Authorization Request (PAR) that included different data rates at different ranges in various radio channel conditions.

IEEE 802.15.3a group was chartered to provide high speed WPAN PHY. The IEEE 802.15.3a PAR described the expectations from IEEE 802.15.3a for the UWB PHY:

- To provide a standard for a low complexity, low cost, low power consumption alternate PHY for 802.15.3.
- To prescribe data rate to be high enough, 110 Mbps or more, to satisfy an evolutionary set of consumer multimedia industry needs for WPAN communications. For example, high data rates, at least 110 Mbps, is required for simultaneous time dependent applications such as multiple HDTV video streams without sacrificing the

requirements for low complexity, low cost, and low power consumption. In addition to at least 110 Mbps data rate, additional data rates, both lower and higher, may be supported.

- To address the requirements to support multimedia data types in multiple compliant co-located systems and also coexistence with other wireless systems in the best interest of users and the industry.

4.3   Design Choices

This section describes two main and most discussed approaches in the UWB PHY layer design: DS-UWB and MB-OFDM based UWB.

4.3.1   Direct Sequence UWB (DS-UWB)

DS-UWB is based on the classical spread spectrum techniques. The transmitter sends low power gaussian shaped pulses spread over a wide bandwidth, which are coherently received at the receiver. Each of the DS-UWB pulses has an extremely short duration. The multi-path effects can usually be ignored due to short pulse duration. The UWB spectrum is divided into 528 MHz sub-bands. The DS-UWB approach uses single-carrier modulation. Historically, it was also referred as impulse radio technology.

Figures 4.1 and 4.2 illustrate DS-UWB transmitter and receiver.

The data to be transmitted is interleaved and scrambled for removing any correlation between adjacent bits. These bits are coded with convolutional coder. Groups of these coded bits are then taken to DS code generator, where their value is used to select a code to transmit. The codes are then fed into the pulse generator, which generates the pulses with a spectral shape close to the desired final shape based on input codes. The resultant train of pulses is

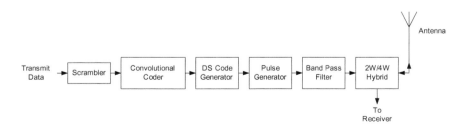

Fig. 4.1 DS-UWB transmitter.

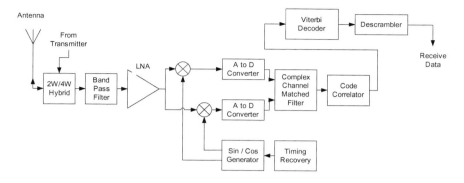

Fig. 4.2 DS-UWB receiver.

fine tuned through a filter. The signal is then fed into a two-wire/four-wire (2W/4W) hybrid, which is connected to the antenna, which finally transmits the signal on air.

Same antenna and hybrid is used by receiver also. At a time, radio is either transmitting or receiving in half duplex mode. Since UWB power levels are very low, the 2W/4W balancing hybrid converts the four-wire transmitter and receiver signals into a two-wire signal that can be connected to an antenna. It uses a balancing network or bridge to make sure that not too much transmit power gets from the transmitter across to the receiver, thereby preventing interference to the receiver of the device by its own transmitter.

As shown in Figure 4.2, the receive signal goes into the hybrid from the antenna. This signal goes to a filter, which removes any out-of-band interference, and then amplified using a Low Noise Amplifier (LNA). The output is demodulated by a quadrature demodulator. The demodulator is fed by a local oscillator. The output of the demodulator is fed into two Analog to Digital Converters (ADC), one each for the real and imaginary demodulator outputs. These outputs go into a complex channel matched filter. The matched filter output is correlated with the code possibilities by code correlator, and the bits are arrived corresponding to the best match. These bits are fed into the Viterbi decoder, and the decoder output is descrambled/deinterleaved to obtain the receive data.

Direct sequence spread spectrum being the historical approach to UWB, DS-UWB was one of the two main proposals for high speed WPAN in IEEE 802.15.3a group, the other proposal being MB-OFDM UWB (explained later in this chapter). It was proposed to be used with IEEE 802.15.3 MAC. Some of the other features of DS-UWB are as below:

- It supports three separate mode of operations, low band (3.1– 5.15 GHz), high band (5.825–10.6 GHz), and multiband (3.1– 5.15 GHz plus 5.285–10.6 GHz).
- Multiple accessing is achieved using Code Division Multiple Access (CDMA).
- Supports for CSMA/CCA, which is helpful in contention based data transfer and useful in situation of simultaneously operating piconets in a spatial region.
- Supports for ranging and location awareness.
- Supports for data rate from 28.5 Mbps to 800 Mbps (can be extended to 1.2 Gbps). Tables 4.1 and 4.2 show DS-UWB rate with information about modulation and CDMA code types.

More details about DS-UWB can be referred in [93] and [101].

4.3.2 Multi-band OFDM UWB (ECMA-368 UWB)

Orthogonal Frequency Division Multiplexing (OFDM) is one of the most popular subcarrier modulation and multiplexing technique for the high data

Table 4.1 Low band DS-UWB rates.

Data rate (Mbps)	Modulation	CDMA Code Type
28.5	BPSK	2-BOK
57	BPSK	4-BOK
75	BPSK	8-BOK
100	BPSK	4-BOK
114	BPSK	8-BOK
200	BPSK	16-BOK
400	QPSK	16-BOK

Table 4.2 High band DS-UWB rates.

Data rate (Mbps)	Modulation	CDMA Code Type
100	BPSK	4-BOK
114	BPSK	4-BOK
200	BPSK	4-BOK
300	BPSK	8-BOK
400	BPSK	16-BOK
600	QPSK	8-BOK
800	QPSK	16-BOK

rate wireless applications. It provides good immunity to multi-path fading and considered superior to CDMA at very high data rates. It is being used in WiMAX (IEEE 802.16 based technologies), DVB-H, while being considered for 3GPP Long Term Evolution (LTE) and 4G technologies.

Multi-band OFDM proposal, another main proposal before IEEE 802.15.3a, was promoted by the WiMedia Alliance and was adopted by ECMA as ECMA-368 standard. This proposal utilizes the unlicensed 3.1 to 10.6 GHz frequency band and supports data rates of 53.3, 80, 106.7, 160, 200, 320, 400, and 480 Mbps. Support for transmitting and receiving data rates of 53.3, 106.7, and 200 Mbps is mandatory.

As shown in Figure 3.1 the UWB spectrum is divided into 14 bands, each with a bandwidth of 528 MHz. The first 12 bands are then grouped into four band groups consisting of three bands. The last two bands are grouped into a fifth band group. A sixth band group is also defined within the spectrum of the first four, consistent with usage within worldwide spectrum regulations. At least one of the band groups must be supported by each UWB device.

Table 4.3 shows the band group allocations with their lower, center, and upper frequency.

Table 4.3 Band group allocations.

Band group	Band ID	Lower frequency (MHz)	Center frequency (MHz)	Upper frequency (MHz)
1	1	3168	3432	3696
	2	3696	3960	4224
	3	4224	4488	4752
2	4	4752	5016	5280
	5	5280	5544	5808
	6	5808	6072	6336
3	7	6336	6600	6864
	8	6864	7128	7392
	9	7392	7656	7920
4	10	7920	8184	8448
	11	8448	8712	8976
	12	8976	9240	9504
5	13	9504	9768	10032
	14	10032	10296	10560
6	9	7392	7656	7920
	10	7920	8184	8448
	11	8448	8712	8976

In the MB-OFDM PHY, a total of 110 subcarriers (100 data subcarriers and 10 guard subcarriers) are used per band to transmit the information. In addition, 12 pilot subcarriers are included for coherent detection. "Frequency Domain Spreading," "Time Domain Spreading," and "Forward Error Correction (FEC) coding" are used to vary the data rates. The FEC used is a convolutional code with coding rates of 1/3, 1/2, 5/8, and 3/4. The coded data is then spread using a Time Frequency Code (TFC).

Unique logical channels are defined by using up to ten different time frequency codes for each band group. The ECMA-368 standard also specifies three types of TFCs:

(i) Where the coded information is interleaved over three bands, referred to as Time Frequency Interleaving (TFI).

(ii) Where the coded information is interleaved over two bands, referred to as two-band TFI or TFI2.

(iii) Where the coded information is transmitted on a single band, referred to as Fixed Frequency Interleaving (FFI).

Support for TFI, TFI2, and FFI is mandatory. Within the first four and the sixth band groups, four time frequency codes using TFI and three time frequency codes using each of TFI2 and FFI are defined; thereby providing support for up to ten channels in each band group. For the fifth band group, two time frequency codes using FFI and one using TFI2 are defined.

A mechanism is provided to allow individual OFDM subcarriers to be nulled, discussed later in this section. This, together with the choice of frequency bands and of TFI, TFI2, and FFI time frequency codes, provides substantial control over the use of spectrum by the transmitted signal, allowing the PHY to be used in a range of regulatory and radio coexistence scenarios.

For the MAC purpose, channelization scheme is defined based on band group and time frequency codes. Channels using TFCs 1–4 are also referred as TFI channels, those using TFCs 5–7 are referred as FFI channels, and channels using TFCs 8–10 are referred to as TFI2 channels.

Table 4.4 shows mapping of channel number from band group and time frequency code.

Table 4.4 Mapping of channel number to band group and TFC.

Channel number	Band group	TFC
9–15	1	1–7
17–23	2	1–7
25–31	3	1–7
33–39	4	1–7
45–46	5	5–6
49–55	6	1–7
72–74	1	8–10
80–82	2	8–10

The PHY layer consists of two protocol functions:

1. A PHY Convergence Function, which adapts the capabilities of the Physical Medium Dependent (PMD) device to the PHY service. This function is supported by the Physical Layer Convergence Protocol (PLCP), which defines a method of mapping the PLCP Service Data Units (PSDU) into a frame format suitable for sending, and receiving user data and management information between two or more UWB devices.

2. A PMD device, whose function defines the characteristics and method of transmitting and receiving data through a wireless medium between two or more UWB devices, each using the ECMA-368 PHY.

Following subsections briefly describes the essential PHY sublayers (PLCP, PMD) and PHY Layer Management Entity (PLME).

PLCP (*Physical Layer Convergence Protocol*) *sublayer*: In order to allow the MAC to operate with minimum dependence on the PMD sublayer, the PHY convergence sublayer is defined. This function simplifies the PHY service interface to the MAC services.

PMD (*Physical Medium Dependent*) *sublayer*: The PMD sublayer provides a mechanism to send and receive data between two or more devices.

PHY Layer Management Entity (*PLME*): The PLME performs management of the local PHY functions in conjunction with the MAC management entity.

Figure 4.3 shows a realization of the transmitted RF signal using three frequency bands. The first symbol is transmitted on a center frequency of 3.432 GHz, the second symbol is transmitted on a center frequency

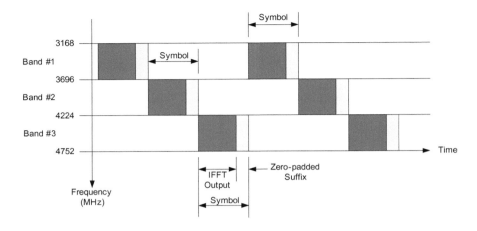

Fig. 4.3 Example — Realization of a transmitted RF signal using three bands.

of 3.960 GHz, the third symbol is transmitted on a center frequency of 4.488 GHz, and the fourth symbol is transmitted on a center frequency of 3.432 GHz, and so on. In addition, it is apparent from Figure 4.3 that the symbol is created by appending a Zero-Padded Suffix (ZPS) to the Inverse Fast Fourier Transform (IFFT) output, or equivalently, to the OFDM symbol. The zero-padded suffix serves two purposes: it provides a mechanism to mitigate the effects of multi-path; and, it provides a time window (a guard interval) to allow sufficient time for the transmitter and receiver to switch between the different center frequencies. A symbol is defined as an OFDM symbol (IFFT output) plus a zero-padded suffix.

4.3.2.1 Tone Nulling

In order to support avoidance of other users of the UWB band, the transmitted signal is sent in the context of a configured array of 384 tone nulling elements. These correspond to the subcarriers of each band within the current band group. Each tone nulling element can take the value one or zero. If the value is zero, then the transmitter should take steps to minimize the transmitted signal energy at the frequency of the corresponding subcarrier. If the value is one, then the signal is unaffected by tone nulling. In ECMA-368 PHY, a device needs to transmit at least 86 useful tones per band, where useful tones relate to tones containing data, pilot, the preamble or the channel estimation sequence.

This limit prevents unacceptable degradation of packet detection performance and other receive performance. If more tones in a band must be avoided, the entire band cannot be used for transmission.

A device may null additional tones beyond those specified, for instance to improve or preserve symmetry within the transmitted symbols, subject to the constraint that at least 86 useful tones must be transmitted per band. If additional tones are nulled then this needs to be done consistently throughout the packet. The simplest possible implementation is to set to zero the corresponding values of IFFT inputs, and to generate the sync symbols through the same IFFT process as other symbols.

4.3.2.2 PLCP Protocol Data Unit (PPDU)

Next we explain the method for converting a PSDU into a PLCP Protocol Data Unit (PSDU). During the transmission, the PSDU is pre-appended with a PLCP preamble and a PLCP header in order to create the PPDU. While at the receiver side, the PLCP preamble and PLCP header are used in the demodulation, decoding, and delivery of the PSDU to MAC.

PLCP Preamble: The goal of the PLCP preamble is to aid the receiver in timing synchronization, carrier-offset recovery, and channel estimation. It consists of packet/frame synchronization sequence and a channel estimation sequence. A unique preamble sequence is required to be assigned to each TFC. The preamble is defined to be a real baseband signal, which is inserted into the real portion of the complex baseband signal. The packet/frame synchronization sequence can be used for packet acquisition and detection, coarse carrier frequency estimation, coarse symbol timing, and for synchronization within the preamble. Whereas, the channel estimation sequence can be used for estimation of the channel frequency response, fine carrier frequency estimation, and fine symbol timing. Preamble is defined as standard preamble or burst preamble. The burst preamble is only used when a burst of frames is transmitted, where frames are separated by a minimum inter frame space time. For data rates of 200 Mbps and lower, all the frames use the standard preamble, though it is burst of data. Device must support for burst preamble if supporting data rate higher than 200 Mbps. The preamble type bit in the PHY Header describes the type of preamble that will be used in the next packet.

PLCP header: The goal of PLCP header is to convey necessary information about both the PHY and the MAC to aid in decoding of the PSDU at the receiver. It consists of PHY header, MAC header, Header Check Sequence (HCS), tail bits, and Reed–Solomon parity bits. Tail bits are added between the PHY header and MAC header, HCS and Reed–Solomon parity bits and at the end of the PLCP header in order to return the convolutional encoder to the "zero state." The resulting scrambled and Reed–Solomon encoded PLCP header is encoded using a $R = 1/3, K = 7$ convolutional code, interleaved using a bit interleaver, mapped onto a Quadrature Phase Shift Keying (QPSK) constellation, and finally, the resulting complex values are loaded onto the data subcarriers for the Inverse Discrete Fourier Transform (IDFT) in order to create the baseband signal. Tone nulling, if implemented, is the applied then.

- *PHY Header*: The PHY Header contains information about the data rate of the MAC frame body, the length of the frame payload (which does not include the Frame Check Sequence (FCS)), the seed identifier for the data scrambler, and information about the next packet — whether it is being sent in burst mode and whether it employs a burst preamble or not. It also includes the information about the TFC and band group used for the frame transmission.
- *Reed–Solomon Outer Code for the PLCP header*: The Reed–Solomon parity bits are added in order to improve the robustness of the PLCP header. The PLCP header uses a systematic (23, 17) Reed–Solomon outer code to improve upon the robustness of the $R = 1/3, K = 7$ inner convolutional code.
- *Header Check Sequence*: The combination of PHY header and the MAC header is protected with a two octet Cyclic Redundancy Check (CRC-16) Header Check Sequence (HCS).

PSDU: This is formed by concatenating the frame payload with the FCS, tail bits, and finally pad bits, which are inserted in order to align the data stream on the boundary of the symbol interleaver. When transmitting the packet, the PLCP preamble is sent first, followed by the PLCP header, and finally by the PSDU. The PSDU is sent at the desired data rate of 53.3, 80, 106.7, 160, 200, 320, 400 or 480 Mbps.

Figure 4.4 shows the standard PPDU structure.

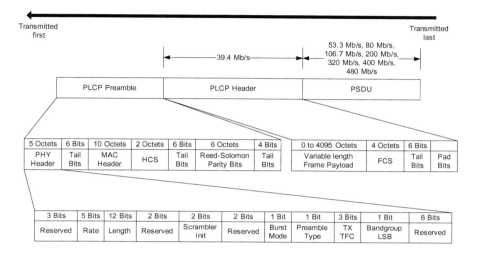

Fig. 4.4 PPDU.

Table 4.5 PSDU data rate dependent modulation parameters.

Data rate (Mbps)	Modulation	Coding rate	Coded bit/ 6 OFDM symbol	Info bits/ 6 OFDM symbols
53.3	QPSK	1/3	300	100
80	QPSK	1/2	300	150
106.7	QPSK	1/3	600	200
160	QPSK	1/2	600	300
200	QPSK	5/8	600	375
320	DCM	1/2	1200	600
400	DCM	5/8	1200	750
480	DCM	3/4	1200	900

Table 4.5 shows the PSDU data rate dependent modulation parameters.

The binary data is mapped onto a QPSK constellation for the data rates 200 Mbps and lower. The binary data is mapped onto a multi-dimensional constellation using a Dual-Carrier Modulation (DCM) technique for data rates 320 Mbps and higher.

Table 4.6 shows some of the timing parameters related to MB-OFDM PHY layer.

OFDM symbols could be easily created using IFFT process. Fast Fourier Transform (FFT) operation could be used at the receiver, which is suitable to be implemented using high speed Digital Signal Processors. This leads to good CMOS realizations.

Table 4.6 Timing parameters.

Parameter description	Value
Sampling frequency	528 MHz
Total number of subcarriers (FFT size)	128
Number of data subcarriers	100
Number of pilot subcarriers	12
Number of guard subcarriers	10
Total number of subcarriers used	122
Subcarrier frequency spacing	4.125 MHz
IFFT and FFT period	242.42 ns
Symbol interval	312.5 ns
Symbol rate	3.2 MHz
Total number of samples per symbol	165

Within the OFDM modulation process, frequency domain spreading within a symbol and time domain spreading across two consecutive symbols is used to obtain further bandwidth expansion, beyond that provided by the forward error correction code and the time frequency codes. Frequency domain spreading entails transmitting the same information (complex number) on two separate subcarriers within the same OFDM symbol. Time domain spreading involves transmitting the same information across two consecutive OFDM symbols. This technique is used to maximize frequency-diversity and to improve the performance in the presence of other noncoordinated devices. Since excellent texts are available on the OFDM, we do not intend to discuss this part in details in this book (Refer [63] and [64]).

4.4 MB-OFDM Transmitter Operation

Transmit data is taken as bit stream by the MB-OFDM transmitter and various operations are carried out through different modules prior to sending bits on air in electromagnetic form. This is described in Figure 4.5. The input bit stream is scrambled. A FEC code is applied through convolutional encoder, which provides protection against transmission errors and data stream can be corrected if there are errors at the receiver end. The encoded sequence is then interleaved (TFI or FFI), mapped to an OFDM symbol (QPSK or DCM) and preamble is added. IFFT transforms the frequency domain information into a time domain OFDM symbol. Finally, Digital to Analog Converter converts the OFDM symbols to time domain analog waveforms, which is converted to the appropriate center frequency and transmitted through the antenna.

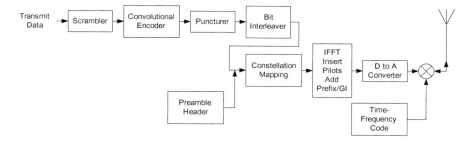

Fig. 4.5 MB-OFDM transmitter.

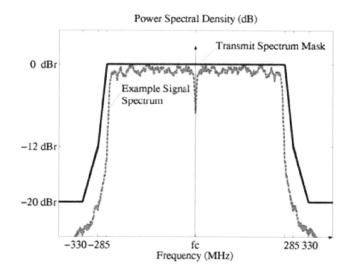

Fig. 4.6 Power spectral density mask for transmitter.

The transmitted spectral density of the transmitted signal needs to fall within the spectral mask, as defined in Figure 4.6.

The transmit center frequencies and the symbol clock frequency are derived from the same reference oscillator with maximum tolerance of $+/-20$ ppm.

Each device provides the support for Transmit Power Control (TPC). The goal of a power control algorithm is to minimize the transmit power spectral density, while still providing a reliable link for the transfer of information. The start of a valid OFDM transmission at a receiver level equal to or greater than the minimum sensitivity for a 53.3 Mbps transmission causes CCA to indicate that the channel is busy with a probability exceeding 90%.

4.5 MB-OFDM Receiver Operation

The received signal at antenna is amplified using a LNA and converted to the complex baseband. Low pass filters remove the out of band interference from this complex baseband signal. The analog signal is converted to complex digital baseband signal by ADC. Baseband processing does the packet detection and FFT operation that obtains the frequency domain information. The output of the FFT is equalized using a Frequency Domain Equalizer (FEQ). A phase correction is applied to the output of the FEQ which can fix the carrier and timing mismatch between transmitter and receiver. The pilot tones in each OFDM symbol drive the digital Phase Locked Loop. The output of the FEQ is de-mapped and de-interleaved which goes to the Viterbi decoder. The error corrected bit sequence from Viterbi decoder is descrambled and passed on to the MAC. Figure 4.7 describes the MB-OFDM receiver pictorially.

Two important considerations in MB-OFDM PHY receivers are receiver sensitivity and receiver Packet Error Rate (PER). In order to ensure a PER of less than 8% with a PSDU of 1024 octets, the minimum receiver sensitivity

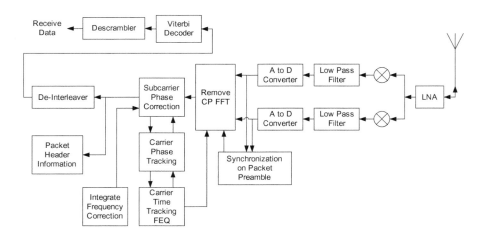

Fig. 4.7 MB-OFDM receiver.

Table 4.7 Minimum receiver sensitivity for band group 1 devices.

Data rate (Mbps)	Minimum receiver sensitivity (dBm)
53.3	−80.8
80	−78.9
106.7	−77.8
160	−75.9
200	−74.5
320	−72.8
400	−71.5
480	−70.4

numbers in Additive White Gaussian Noise for the different data rates are listed in Table 4.7.

Here a noise figure of 6.6 dB (referenced at the antenna), an implementation loss of 2.5 dB, and a margin of 3 dB is assumed. These sensitivity numbers are prone to variations in noise figure over process, voltage, and temperature.

Further, the device needs to estimate the link quality of the received channel, where the link quality is defined as an estimate of the Signal to Noise Ratio available. A device may indicate the strength of the receiving signal in decibels. Receive Signal Strength Indicator (RSSI) is a function of the energy received at the antenna.

4.6 Ranging and Location Considerations

The ECMA-368 standard provides an optional feature for a device to support ranging. A device can support the capability to determine the round trip delay between itself and another device. This method is also called Two Way Time Transfer (TWTT) method. This round trip delay is used to calculate the distance between the devices. The distance can be estimated by multiplying the speed of light by the measured propagation delay between the devices.

Ranging is supported with an accuracy and precision of ±60 cm or better. This is important as UWB is a short range technology and ranging accuracy less than this has no meaning.

Ranging operation is performed with the help of MAC ranging commands and PHY ranging support. Ranging support requires at least one ranging timer clocked at 528 MHz. Delay parameters have to be taken into consideration by MAC for time measurement for ranging through ranging timers.

4.7 Coexistence with Other Technologies: Detect and Avoid Mechanisms

Detect and Avoid (DAA) is one of the techniques for solving the interference to licensed spectrum users (e.g. WiMAX) due to UWB transmissions. This technology protects the detected victim device against potential harmful interference, while allowing the same spectrum for UWB devices. Rules for DAA emerged under European Radio Communication Office (ERO), Electronic Communications Committee (ECC), and Task Group 3 for UWB. Other regions including Europe have different licensed users of different narrowband spectrums under UWB spectrum, which prompted UWB developers and promoters to consider DAA mechanisms in ECMA-368 PHY.

This approach allows the UWB transmitter to (a) detect another active service and likelihood of UWB transmitter causing high interference to the other service, (b) avoids the interference to other service by reducing its own radiated power (e.g. using transmit power control) in the specific band. Therefore, the DAA allows UWB systems to increase its maximum Effective Isotropic Radiated Power (EIRP) when it does not detect presence of other radio services and avoid transmitting in the victim service bands when a victim signal is detected. This allows UWB systems to operate across a continuous wide spectrum band. Key to successful UWB operation lies in specifying DAA mechanisms in its devices, keeping in mind current and forecasted active services in the spectrum band where UWB will operate.

Three operation zones are defined for the UWB device depending on the device's proximity upon detection of the victim device. In simplistic term, UWB device can avoid causing interference to victim device by reducing signal strength of the transmission, on detection of the victim device. Figure 4.8 shows DAA zone structure pictorially. If distance between victim device and UWB device is equal or less than 3.5 m, then UWB device is in zone one (UWB device is very near to victim). As distance between victim and UWB device increases, the zone number increases. Zone one being nearest to victim, requires highest protection by the UWB devices in that zone. Highest protection means lowest signal strength from the UWB device. Less protection is required when UWB device is further apart from the victim, means signal strength of UWB device can be higher than signal strength in zone one.

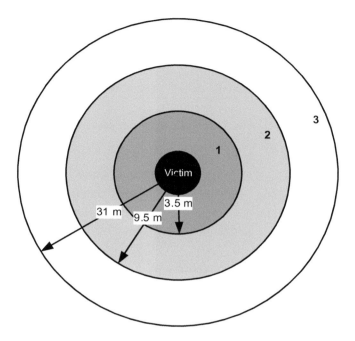

Fig. 4.8 DAA zones.

Multiple zones allow UWB devices to adapt to avoid interference yet allow as much signal strength as possible.

Typically, OFDM allows for effective DAA mechanisms compared to impulse radio UWB systems, through "nulling" specific subcarriers. In fact, a flexible DAA approach may be defined, which is based on combination of victim services down-link and up-link detection, allowing a real coexistence mechanism between UWB devices and victim services. Simple energy detection can be used in UWB chips to detect either down-link or up-link signals from victim.

Down-link detection is based on detection of signal coming from base station. These signals are generally with high signal strength and transmitted continuously. Down-link confirms the victim service (e.g. WiMAX base station) detection. For protection of near-by victim device, up-link detection detects the signal strength of the up-link transmission of the victim device (e.g. WiMAX device). If required then protection is employed. Zone information can be deducted from signal strength detected. With combination of down-link and up-link detection DAA procedure can be summarized as follows:

- Down-link detection to detect victim service deployment, e.g., broadcast mode.
- Up-link detection to detect client victim service and DAA zone.
- Avoid operation based on different maximum UWB emission level.

To eliminate the interference UWB device reduces the signal strength in same frequency band. There are several other methods that can be used for interference avoidance depending on the UWB device's DAA zone, e.g. changing to a different channel, using Low Duty Cycle, creating special notch.

DAA techniques are under validation, jointly carried out by WiMedia Alliance and WiMAX Forum, so it can be defined and used widely in interest of UWB device and victim services.

4.8 Summary

In this chapter, we discussed two main PHY layer approaches, one based on Direct Sequence UWB and the other based on Multi-band OFDM. In particular, we emphasized Multi-band OFDM based PHY layer which has been adopted by ECMA-368. We also discussed briefly on the ranging support. DAA mechanism is introduced that allow flexibility in varied regulatory constraints.

5

Medium Access Layer

5.1 Introduction

Support for various high data rate applications and unique requirements of WPAN are the driving factors for Ultra Wideband (UWB) systems. To fulfill same, the Medium Access Control (MAC) protocol for UWB systems is specified as ECMA-368. We will discuss how expectations from new generation of short range wireless can be met with various MAC techniques. In this chapter and following chapters, we explain the MAC principles for the UWB systems.

5.2 Requirements

Basic requirement for any wireless MAC is to provide the system for transfer of data on broadcast wireless medium. Till today, numerous wireless PHY and MAC technologies have been defined and implemented providing various mechanisms for achieving data communication.

Wireless is a shared medium, multiple devices active in one physical channel need to share the channel for communication. Using channel sharing mechanism is generally referred as medium access. Some of the well known medium access methods for wireless are Time Division Multiple Access (TDMA), Carrier Sense Multiple Access (CSMA), Code Division Multiple Access etc. Combinations of multiple mechanisms have also been used depending on capabilities of the system and requirements of applications.

In addition to medium access, MAC also provides the mechanisms for device discovery, timing synchronization, channel sensing, channel access, data transfer, power save, security, and support for upper layer protocol.

One important requirement for wireless MAC for handheld devices is the capability to support mobility of the device. This can be done in centralized, decentralized, or other topology depending on the requirements.

Following subsections discuss requirements of UWB wireless system from MAC perspective.

5.2.1 Device Discovery

Device can transfer the data to another device in radio range. Device needs to know about the presence of the neighbor before any communication. Besides this, device has to know about the capabilities of the neighbor before the data transfer to ensure optimal operation. These requirements can be fulfilled by device discovery.

5.2.2 Timing Synchronization

Two physically separate devices always have some clock timing difference. Two devices must have same view of time in order to communicate with each other. Due to difference in clocks, it is impossible to schedule transmission and reception with proper timings accuracy. Without synchronized clocks, multiple devices' attempt to use wireless medium may result in collision. Any coordinated power save mechanisms would also be difficult to achieve.

5.2.3 Data Transfer

Being high data rate wireless system, UWB should support different types of traffic for wider acceptability. It should support isochronous traffic with constant service interval (e.g. video, audio, HDTV stream, etc.); besides supporting asynchronous and bursty traffic for applications like internet browsing.

Wireless communication systems for handheld and mobile devices also have limited buffer size, so it is expected to support different kinds of traffic in limited buffer size with tight service interval requirements.

Multicast at wireless MAC layer is a ubiquitous feature, which provides high level of spatial reuse for a set of devices. Many legacy MACs do not support multicasting. This feature brings high efficiency to the applications requiring multicast support.

5.2.4 Power Save

Most of the wireless devices are mobile and battery operated. These devices cannot have power cable attached along and battery is a very costly resource for any electronics system. Any saving in power can extend the life of the battery and make the device long lasting, which eventually leads to better user experience.

5.2.5 Security

Wireless systems are prone to attacks because of their broadcast nature. Not applying encryption methodologies ultimately results into clear text transmission. One can easily steal identity of someone else in network and communicate on behalf of it. The whole communication conversation can be hijacked and can be resumed later at any time for retrieval of confidential information. Without security implementation, alteration of transmitting information can be done easily with Man-In-Middle attacks being a common threat.

5.2.6 Support for Upper Layers

Unlike many legacy MAC systems which are tightly coupled to a single upper layer protocol stack, UWB common radio architecture is expected to support multiple upper layer protocols. This helps to support already existing large application base.

Basic frame format and channel access mechanism are the base of MAC architecture. Upper layer protocol can be adapted through different adaptation layers depending on requirements of the upper layer. Multiple upper layer protocols can use single common radio link.

5.3 Design Choices

In following subsections, we discuss different designs for MAC protocol with their advantages and disadvantages. ECMA-368 MAC chose some of these techniques considering UWB system requirements from the possibilities and diverse protocol choices.

5.3.1 Centralized Vs Decentralized Network Topology

Many of the legacy MAC layers have centralized network topology, where a coordinator manages whole network. Devices other than coordinator, in radio range of the coordinator, associate with coordinator and use the services provided by coordinator. So, centralized network has basically two types of devices, coordinator and devices.

In wireless personal area networks, network with one coordinator and many associated devices is called piconet. The coordinator provides many complex MAC functionalities and serves devices in the piconet, so it is complex to implement. The coordinator usually has large memory buffers and more power or connected to power line. Features of coordinator make coordinator's mobility limited and that limits the mobility of connected devices. Figure 5.1 depicts piconet with coordinator.

The device of centralized network is simple and easy to implement, because complexity is in the coordinator. Disadvantage is that, device's functionalities are too much dependant on coordinator. Breakdown of coordinator can jeopardize whole network. Simultaneously operating piconets in range of each other can create interference due to hidden node problem. Figure 5.2 shows simultaneously operating piconets. If a device goes out of range of one coordinator due to mobility and becomes part of another piconet, then it involves complex method of hand-off and service continuity.

On other hand, decentralized network does not have any coordinator and each device is self contained. In absence of a coordinator, device has to support

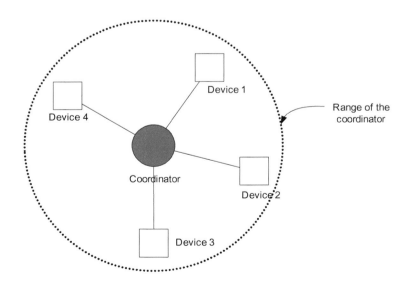

Fig. 5.1 Centralized network (Piconet).

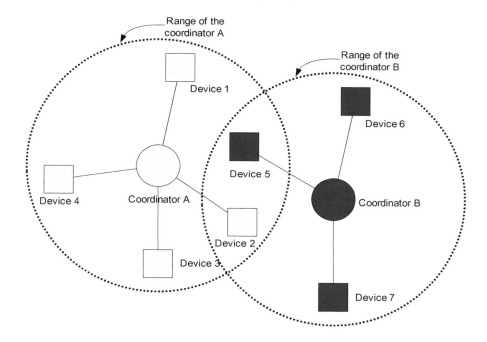

Fig. 5.2 Simultaneously operating piconets.

all MAC services by itself. Device takes care of itself and its neighborhood. Device can communicate with any other device in neighborhood (in radio range). Support of many MAC functionalities makes this device complex to implement. Instead of piconet, the topology of decentralized network is like cloud of devices, shown in Figure 5.3.

Decentralized topology is better suited to support mobility of the device. Device is not dependent on anybody other than the devices with whom it communicates. Multiple piconets in range can operate since it is cloud of devices. This network architecture provides basic MAC functionality like channel access, data transfer, and power save. Each device has same MAC functionality so it is easy to adapt it to multiple upper layer protocols. It provides the spatial reuse for data and control traffic as shown in Figure 5.4.

Depending upon the use case and system requirements, either centralized or decentralized approach could be adopted. The WPAN systems based on

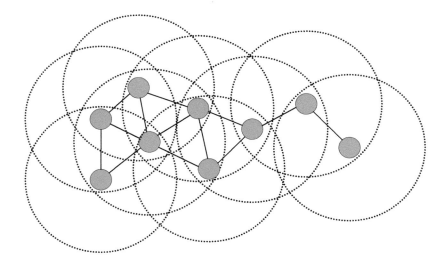

Fig. 5.3 Cloud of devices.

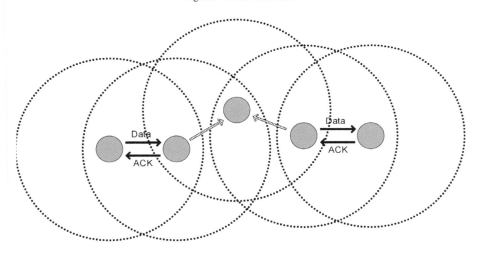

Fig. 5.4 Spatial reuse.

IEEE 802.15.1 (for Bluetooth) and IEEE 802.15.3b employ the centralized MAC approach.

Support of mobility and spatial reuse make the decentralized approach very appealing to the requirements of UWB system, hence it was adopted in the ECMA-368 MAC.

5.3.2 Device Discovery

Beacon is the frame generally used for device discovery. Beaconing can even be used for protocol negotiations in distributed and decentralized network. Beacon must not collide and reach to all the neighbors, otherwise it can be difficult to discover the existence of the device.

Beacons can be transmitted randomly with random time interval. Receiving devices do not know when the beacons are going to come. This forces all devices to be awake all the time to receive beacons appearing on-air any time. As devices have to be awake to listen to randomly transmitted beacons, possibilities of saving power are compromised. Further, random interval between beacons leads to noncontinuous time intervals for data transfer. Random beacons do not guarantee the reception of the beacon by each neighbor devices, because multiple random beacons from different devices can be transmitted at the same time and can overlap and collide. However, it is easy to implement and involves no complex protocol. Figure 5.5 shows random beacons in the air.

Another way to broadcast beacons is to distribute the beacons at constant time interval, as shown in Figure 5.6. Different devices in the network transmit the beacon such that time interval between two beacons remain constant. It is comparatively complex to implement than random beacons. This method also does not allow long continuous time interval for data transfer. Devices have to be awake frequently for reception of distributed beacons by different devices. This method hinders the applications with requirement of large continuous data transfer period.

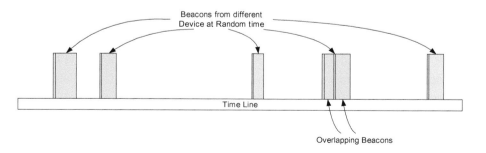

Fig. 5.5 Random beacons.

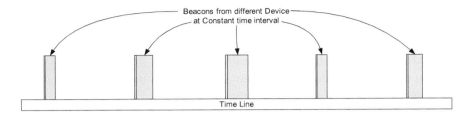

Fig. 5.6 Distributed beacons.

Fig. 5.7 Single beacon period.

Another alternative is transmission of beacon in a Beacon period (BP). BP is time duration at constant interval where all devices transmit a beacon, very near to each other in time. Each device has to be awake during BP to transmit and receive the beacons. If there is no data transfer scheduled then device can sleep for rest of the time. Duration between two BPs can be used for data transfer. Figure 5.7 shows single BP technique for beacon broadcast. It enables long time interval for data transfer and enables power save with long sleep durations. It involves comparatively complex to implement protocol than previous two methods. Service interval requirement of some data stream can be compromised due to large BP (result of many devices).

A combination of two methods, single BP and distributed beacon method (random or distributed beacon), can provide better service internal and power save. This is referred to as multiple BPs. However, this method includes very complex protocol for synchronization and data transfer. Figure 5.8 shows an example of multiple BPs.

Single BP, with variable length of BP, at constant interval provides good power save possibilities and long continuous time for data transfer. In presence of limited number of devices, BP length is small so good service interval can be achieved. Even in worst case, for large number of devices, service interval

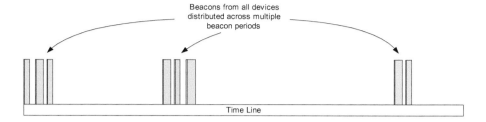

Fig. 5.8 Multiple beacon periods.

achieved is good enough for most of the considered upper layer protocols like wireless USB and internet protocol. These advantages made it best candidate for UWB MAC.

5.3.3 Channel Access Methods

Ultra Wideband technology is expected to cater to applications with different traffic requirements, so MAC must support different types of data transfer to meet these requirements. Further, MAC has to support multiple simultaneously operating upper layers with different active multiple communications. Depending on support provided by UWB PHY, here we discuss different access methods possible for MAC.

Time Division Multiple Access divides continuous time in multiple small time intervals. These time intervals are used by different devices for data transfer. It does not require any Clear Channel Assessment (CCA) before transmission because the time duration is reserved for the device's transmission. Data transfer can begin right at the start of time interval. TDMA provides great utilization of channel time. This method can support isochronous traffic very well because of reserved time slots for communication. It is inefficient for bursty traffic where traffic generation is not constant. If TDMA is used for bursty traffic (e.g. IP traffic through internet browsing) then it is possible that allocated time interval is not utilized fully because of bursty nature of the traffic. Figure 5.9 illustrates TDMA slots reserved for transmission and reception.

Carrier Sense Multiple Access — Collision Avoidance a traditional method used in IEEE 802.11, requires channel sensing and back off before transmission. Being a best effort method, it does not guarantee on time transmission of

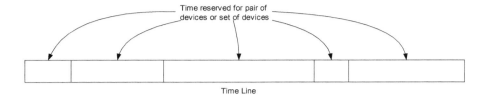

Fig. 5.9 TDMA slots.

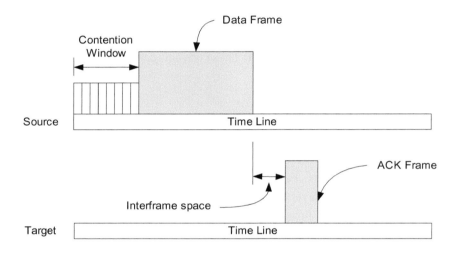

Fig. 5.10 CSMA/CA.

the frame. Here, the medium is shared amongst the devices present in the radio range of the transmitting device. This method is not suitable to isochronous traffic because it does not provide guarantee on time of transmission. But, it is suitable for bursty traffic because medium is shared and can be used by another device if it is unused. Figure 5.10 displays how CSMA/CA is used for data transfer.

Ultra Wideband MAC includes combination of both the methods to cater multiple applications with different type of traffic requirements. It allows reservation of time slot based on TDMA for isochronous traffic and unreserved time is used by variant of CSMA/CA as suitable medium for bursty traffic. Details on data transfer are explained in this chapter later.

5.3.4 Power Save

Ultra Wideband is being targeted for wireless mobile, battery operated and handheld devices so power saving is essential. Power save either can be for long time like several milliseconds or can be for short time in microseconds.

Long power saves are generally driven by application or platform which is using the wireless connectivity, e.g. device is in standby mode, application is in inactive period, device suspends its operation, etc. During long power save, device does not wakeup to receive neighbor's beacons and traffic. Sleeping device is not known by new devices coming into the network. It also looses the time synchronization and information about its neighborhood while being in long power save. In this case, other devices can buffer the traffic or drop the traffic for the device in long power save. Figure 5.11 shows long power save, while Figure 5.12 shows short power save.

During short power save, devices listen for beacons and know about possible traffic for transmission and reception involving itself. If there is no traffic scheduled then it can sleep during that idle time and has to wakeup again during BP to transmit and receive the beacons. Short power save provides relatively

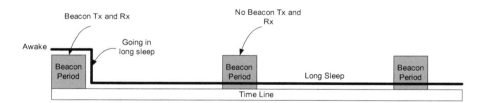

Fig. 5.11 Long power save.

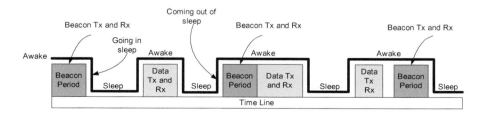

Fig. 5.12 Short power save.

low saving of power but it keeps device active in the network, therefore it does not loose the synchronization and normal operation.

Both the powers save mechanisms have their own advantages and are suitable for UWB system, so both are included in the MAC architecture and used according to suitability.

5.3.5 Security

Security has been a prominent problem for wireless network. There have been constant attacks on existing wireless networks. And security is either not good enough or very difficult to use, so it is not used in deployed systems.

Medium Access Control provides the communication amongst only secure devices, only nonsecure devices or combination of both depending on application requirement.

128 bit Advanced Encryption Standard (AES), which is an approved standard by the US government is adopted for UWB MAC under ECMA-368. AES and CCM are used for data encryption and integrity. It is also quick enough to handle the high speed UWB PHY. Encryption and decryption has to occur in hardware at the full line speed of 480 Mbps, with no significant overhead or delay. Dynamic key generation through 4-way handshakes is done, which can be used for encryption and authentication for pair of devices or group of devices.

5.4 Distributed MAC

Many traditional wireless MAC layers are centralized and driven by coordinators for specific application or upper layer. Proprietary extensions do exist for adaptation of different upper layers, though the MAC may not be designed for that purpose. To cater different applications and upper layers on common radio and considering different design choices as explained in previous subsections, distributed MAC architecture was adopted for UWB system.

Each device is self contained and takes care of itself and its neighborhood in distributed MAC. Data transfer is TDMA (Time slot) based where different types of channel access are handled in the time slots. It also provides Power save features in distributed fashion. Details of distributed UWB MAC are explained in the following subsections.

5.4.1 Network Topology

Figure 5.13 shows the network topology of UWB devices. Circle around the device is radio range of the device. Each device broadcasts its beacon (always unsecured). Signal reach-ability of device's beacon at base lowest data rate defines the radio range of the device. Device can communicate to other device if it is in radio range (distance is less or same as the radio range). Devices which are in radio range of each other are neighbors of each other. Many neighbor devices (cloud of devices) form a network or cluster of devices.

Data (secured or unsecured) can be transferred at same base data rate or higher. At higher data rate, signal reach-ability reduces.

This MAC does not provide multi-hop data transfer or routing functionality.

5.4.2 Addressing

Generally each UWB device has its unique EUI-48 (Extended Unique Identifier) of length 48 bits, popularly known as MAC address. For MAC purposes, the device is identified through 16 bit (shorter than EUI-48) device

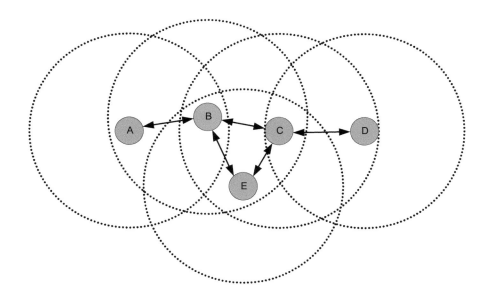

Fig. 5.13 Topology.

address (DevAddr) generated randomly and locally by the device. Due to local and random generation, device address conflict is possible, because the same device address can be generated and used by some other device. Device generates new random device address for its usage in case of conflict.

Device address is used in MAC frames as source address (Device address of transmitting device) and target address (Device address of intended recipient). 16 bit device addresses are divided in four different ranges depending on their usage as shown in Table 5.1.

Private Device Address: It is a reserved class of device address for Application Specific (AS) usage or used by upper layer. Private device address is not randomly generated. Even, device address conflict resolution for private device address is not handled by MAC, but it is the responsibility of upper layers or user of this address needs to make sure that there is no conflict of address during usage. Private addresses are only used for data communication during private reservation.

Generated Device Address: It is randomly generated unicast device address.

Multicast Device Address: This address is used to target group of neighbor devices. One multicast address includes group of devices that are part of multicast group. MAC provides a way to bind a Multicast Device Address to EUI-48 address.

Broadcast Device Address: This address is used to target all neighbor devices.

Source and destination address for the frame for transmission are set as follows:

- Source address of the frame is device's own device address, which could be either generated or private address.
- Destination address for unicast frame is either generated or private address.
- Destination address for multicast frame is multicast address.

Table 5.1 Device address ranges.

Address type	Range
Private	0x0000 - 0x00FF
Generated	0x0100 - 0xFEFF
Multicast	0xFF00 - 0xFFFE
Broadcast	0xFFFF

- Destination address for broadcast fame is set to 0xFFFF, which is broadcast address.

5.4.3 Superframe

The timing structure for UWB MAC is called superframe. The superframe is of $65536\,\mu$s duration. It is divided into 256 equal duration time slots called Medium Access Slot (MAS) each of $256\,\mu$s. Figure 5.14 shows the superframe structure.

Superframe is divided into BP and Data Period (DP). It starts with variable length BP, which extends one or more continuous MASs for transmission and reception of beacons. Remaining part of the superframe is used for data communication, hence called data period. The start of the superframe, is called Beacon Period Start Time (BPST). Internally for the device BPST could be time zero for its reference.

Superframe is also described in the form of allocation zone. Superframe is split into 16 zones numbered from 0 to 15 starting from BPST. Each zone contains 16 consecutive MASs, which are numbered from 0 to 15 within the

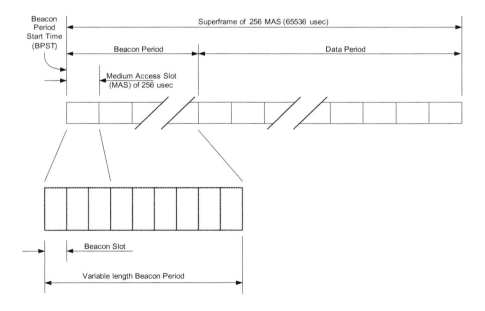

Fig. 5.14 Superframe.

Beacon Zone / Allocation Zone

	0	1	2	3	4	5	6	7	8	9	10	11	12	13	14	15
0	0	16	32	48	64	80	96	112	128	144	160	176	192	208	224	240
1	1	17	33	49	65	81	97	113	129	145	161	177	193	209	225	241
2	2	18	34	50	66	82	98	114	130	146	162	178	194	210	226	242
3	3	19	35	51	67	83	99	115	131	147	163	179	195	211	227	243
4	4	20	36	52	68	84	100	116	132	148	164	180	196	212	228	244
5	5	21	37	53	69	85	101	117	133	149	165	181	197	213	229	245
6	6	22	38	54	70	86	102	118	134	150	166	182	198	214	230	246
7	7	23	39	55	71	87	103	119	135	151	167	183	199	215	231	247
8	8	24	40	56	72	88	104	120	136	152	168	184	200	216	232	248
9	9	25	41	57	73	89	105	121	137	153	169	185	201	217	233	249
10	10	26	42	58	74	90	106	122	138	154	170	186	202	218	234	250
11	11	27	43	59	75	91	107	123	139	155	171	187	203	219	235	251
12	12	28	44	60	76	92	108	124	140	156	172	188	204	220	236	252
13	13	29	45	61	77	92	109	125	141	157	173	189	205	221	237	253
14	14	30	46	62	78	94	110	126	142	158	174	190	206	222	238	254
15	15	31	47	63	79	95	111	127	143	159	175	191	207	223	239	255

Fig. 5.15 Superframe zone structure.

zone. Figure 5.15 illustrates the zone structure of the superframe, also called as MAS map table.

5.4.4   MAC Frame

Figure 5.16 displays PHY frame with MAC frame inside it including transmission order of the fields. PHY frame starts with preamble which carries timing information, followed by PHY header which includes payload length information. MAC header is of 10 octets following PHY header. Header Check Sequence protects PHY and MAC header from being changed.

Medium Access Control header indicates if frame is secured or unsecured. Unsecured frame can accommodate maximally 4095 octets, while secure framed can accommodate maximally 4075 octets as payload. Secured frame also includes security fields like Temporal Key Identifier (TKID), Encryption

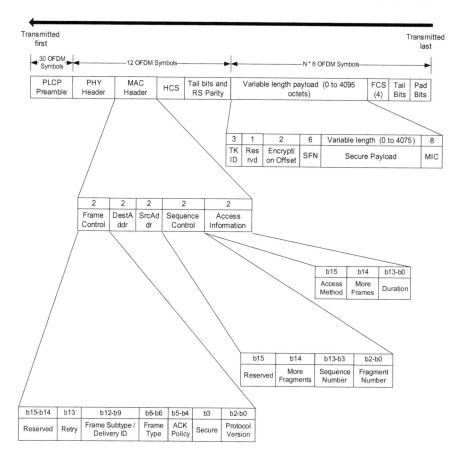

Fig. 5.16 MAC frame format.

Offset (EO), Secure Frame Number (SFN) and Message Integrity Code (MIC), totally 20 octets.

If frame is a secured frame, it is protected by AES for authentication and encryption. EO field indicates that from where the payload is encrypted, so it is possible that secured frame may have some part of payload which is not encrypted but only authenticated. MIC provides authentication. SFN provides freshness and replay prevention. TKID identifies the secure key used for encryption and authentication.

Payload is always followed by 4 octet Frame Check Sequence which protects payload part of the frame against changes and corruption.

Wireless medium is a broadcast medium so each frame is received by all active devices in radio range of the transmitting device. Source address and destination address in MAC header is used to filter the relevant frame and know the identity of the frame origin.

Due to unreliability of the wireless medium frame can be missed, frames may collide, so retry is required, and this is indicated by retry field of MAC header. Frame can also be fragmented to improve the reliability over wireless medium and handle large size. Sequence control field takes care of missed frames and used in reassembly of fragmented frames.

Medium Access Control supports different acknowledgment policies depending on requirement of the traffic. It is indicated in acknowledgment policy field of MAC header.

Access method field is used to identify the access method used for the traffic and invalid traffic can be filtered using this.

Duration field is used in virtual carrier sensing and indicates medium busy, so other devices do not access the medium during duration time for transmission. This enables power saving for other devices.

More frame field is used to indicate that there are more frames to be transmitted. If there are no frame for transmission then this field is set to zero so target device can go in power save.

Different frame types and subtypes are indicated in MAC header. There are six frame types and these frames have subtypes as follows:

- Beacon frame is used by beacon protocol. Beacon frame is heart of MAC control mechanism. It includes different Information Elements (IEs) for different protocol functions.
- Control frame is used for different purposes like acknowledgement, virtual carrier sensing, reservation release, and for application specific usage.
- Command frame can be used for reservation negotiation, probe, security, and range measurement etc. Command frame also can be used for application specific requirement by defining its payload.
- Data frame is used to transfer actual upper layer data. Similarly, aggregated data frame is multiple upper layer data frames in single frame.

Tables below provide brief information about the usage of different frames, Table 5.2 — Control Frames, Table 5.3 — Command Frames, and Table 5.4 — IEs.

5.4.5 Guard Time

Device is required to have clock accuracy of 20 ppm. All time measurement is expected to be with 1 μs resolution to maintain the synchronization, beacon slot boundary, MAS boundary, and frame reception.

Though required resolution of clock is defined, physical difference in clocks, implementations, and synchronization drifts may lead to the situation where reservation boundaries and slot boundaries are not synchronized. Short Inter Frame Space (SIFS) intervals have to be maintained between frames for proper reception of the frames. SIFS interval and guard interval are maintained at the end of each reservation block. SIFS provides enough time for PHY to go to reception mode and be ready for reception of a frame from worst case situation (being in transmit mode). Guard time protects traffic from two

Table 5.2 Control frames.

Control frame	Usage description
Immediate-ACK (Imm-ACK)	Acknowledges the previously received frame with ACK policy set to Imm-ACK
Block-ACK (B-ACK)	Acknowledges the previously received block of frames (multiple frames)
Request to Send (RTS)	Announces to the recipient that frame is ready for transmission
Clear to Send (CTS)	Announces the transmitter that device is ready to receive, send in response to RTS
Unused DRP Reservation Announcement (UDA)	Announcement by the owner of the reservation to neighbors that unused reservation is available for use by neighbors
Unused DRP Reservation Response (UDR)	Announcement by the target of the reservation to neighbors that unused reservation is available for use by neighbors. Sent in response to UDA
AS	Defined for AS usage

Table 5.3 Command frames.

Command frame	Usage description
DRP Reservation Request	Used for DRP reservation negotiation
DRP Reservation Response	Used for DRP reservation negotiation
Probe	Used in probing IEs
Pair-wise Temporal key (PTK)	Used to derive PTK using 4-way handshake
Group Temporal key (GTK)	Used to distribute GTK to group of devices
Range Measurement	Used to exchange timing information during Range measurement
AS	Used for AS purpose

Table 5.4 Information elements.

Information element	Usage description
Traffic Indication Map (TIM) IE	Used by transmitter device to indicate the availability of buffered traffic for transmission
Beacon Period Occupancy (BPO) IE	Indicates the neighbors' BP occupancy in the previous superframe
PCA Availability IE	Indicates the device's availability to receive PCA traffic
DRP Availability IE	Indicates device's availability for new DRP reservation
Distributed Reservation Protocol (DRP) IE	Used for creation and maintenance of DRP reservation
Hibernation Mode IE	Used for hibernation process to indicate the hibernation duration
BP Switch IE	Used for BP switching with information about new BP
MAC Capabilities IE	Used for broadcasting MAC Capabilities
PHY Capabilities IE	Used for broadcasting PHY Capabilities
Probe IE	Used for getting some IEs from other device
AS Probe IE	Used for probing AS IEs
Link Feedback IE	Provides data rate and power related feedback
Hibernation Anchor IE	Provides information about hibernating devices
Channel Change IE	Used for channel changing
Identification IE	Provides information about the device
Master Key Identifier IE	Provides information about the master key(s) held by the device
Relinquish request IE	Used for requesting release of the reservation
Multicast Address Binding IE	Indicates an address binding between a multicast EUI-48 and a multicast device address
Tone-nulling IE	Announces tone-nulling information for Detect and Avoid operation
Regulatory Domain IE	Announces regulatory domain information
AS IE	Used for AS usage

reservation blocks from overlapping and ensures proper reception. Figure 5.17 and calculation below explains the guard time and how the value of guard time is derived.

Guard time is possible maximum absolute difference between two devices MAS boundary.

Guard time = maximum synchronization error + maximum drift.

Maximum synchronization error due to two devices is double of clock resolution.

$$\text{Maximum synchronization error} = 2 \times \text{ clock resolution}$$
$$= 2 \times 1$$
$$= 2\,\mu s$$

Maximum drift includes drift due to maximum clock accuracy of 2 devices and maximum time elapsed after synchronization done.

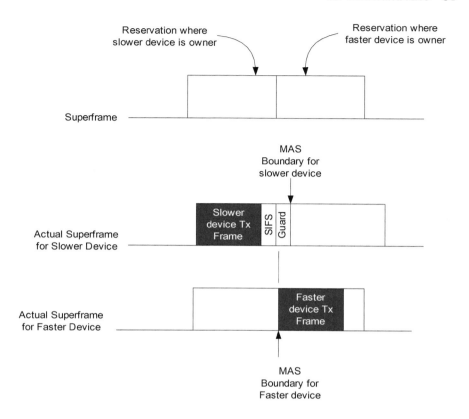

Fig. 5.17 Guard time.

$$\text{Maximum drift} = 2 \times \text{clock accuracy (ppm)}$$
$$\times 10^{-6} \times \text{synchronization interval}$$

Worst case synchronization interval is duration of 4 superframe, i.e. device maintain the history of a neighbor. Propagation delay around 33 ns for such short range is ignored for calculation.

$$\text{Maximum drift} = 2 \times \text{clock accuracy (ppm)}$$
$$\times 10^{-6} \times 4 \times \text{superframe duration}$$
$$= 2 \times 20 \times 10^{-6} \times 4 \times 65536$$
$$= 10.48576 \, \mu s$$

$$\text{Guard time} = 2 + 10.48$$
$$= 12.48\,\mu s$$
$$\approx 12\,\mu s$$

5.5   Beacon Protocol

Each device periodically broadcasts a beacon frame for the purpose of device discovery, capabilities announcements, and control information distribution. Beacon is also a tool for reservation establishment, traffic indication, power save, channel availability information distribution, etc. It includes set of IEs to perform different MAC functions.

Beacon is transmitted during BP and the device receives beacons from neighbors. Device receives beacon from all its neighbors (synchronized with the device) and set of those neighbors form a Beacon Group (BG). The union of a device's BG and the BGs of all devices in the device's BG is called extended BG. Figure 5.18 shows the concept of BG and Extended BG or 2 hop neighborhood.

5.5.1   Beacon Period (BP)

Superframe starts with variable length BP. BP consists time slots of $85\,\mu s$ (which is approximately 1/3 of MAS duration) called beacon slots. Depending on the number of devices in extended BG, BP length could go up to 96 beacon slots. First two beacon slots in BP, called signaling slots, are used for extending BP length. Device transmits beacon in selected beacon slot at the start of the beacon slot. Figure 5.19 shows the typical BP.

Generally, any information received from neighbor devices or events are considered to be valid for three superframes, so any beacon slot is considered available if there is no beacon in extended neighborhood for the last four superframes.

5.5.2   Beacon Period Occupancy IE

Beacon Period Occupancy IE (BPO IE) is an important part of distributed MAC protocol. Device includes BPO IE in each of its beacons. BPO IE provides the information of beacon slot occupancy by neighbor devices. It includes the

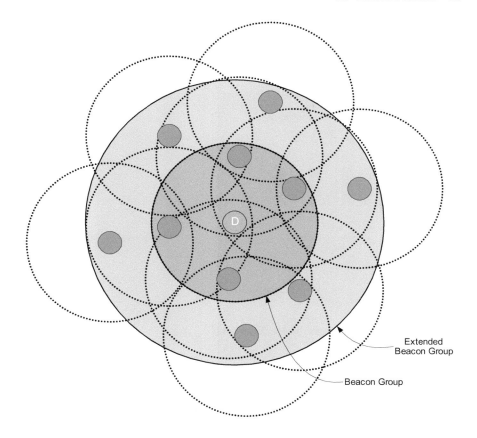

Fig. 5.18 Beacon group and extended beacon group.

list of devices present in its neighborhood in previous superframe including beacon slot and device address of those devices.

Received BPO IEs from the neighbors provide information about the extended BG (2 hop neighborhood) and unused beacon slots in 2 hop neighborhood.

Figure 5.20 shows how BPO IE information is used and topology information can be passed to 2 hop neighborhood. Device O has 3 neighbors. Using the information of those 3 neighbors' BPO IE, O knows that slot up to 8 are occupied in extended BG. Also devices more than 2 hops away can have same beacon slot without any conflict, e.g. Device G and D.

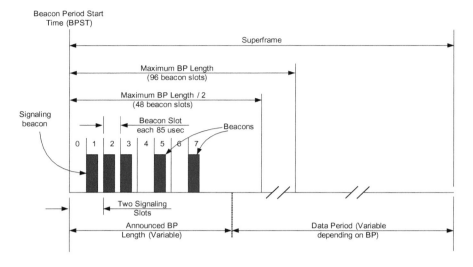

Fig. 5.19 Beacon period.

5.5.3   Beacon Procedure State Machine

Figure 5.21 shows beacon protocol state machine in brief. Device starts the operation by scanning for the beacons in the channel. If any beacon is received in the channel then device synchronizes to the received beacon and joins the BP. If no valid beacon (beacon without any errors) is received then device can start its own BP or can scan for beacons one more superframe. Device can also decide to stop the operation instead of starting its own BP.

5.5.4   Join BP or Start BP

Device transmits beacon in slot number 2 according to its internal clock reference to start new BP. This beacon's BPO IE does not include any neighbor device because of no neighbor present and just includes the information about BP length.

Device synchronizes its clock to received beacons' BP to join the BP. Device selects beacon slot after highest unavailable beacon slot it observed during the scan using BPO IE information of neighbors. It can select slot in 8 slots after highest unavailable. The device transmits beacon in selected beacon slot with BPO IE including neighbors' information.

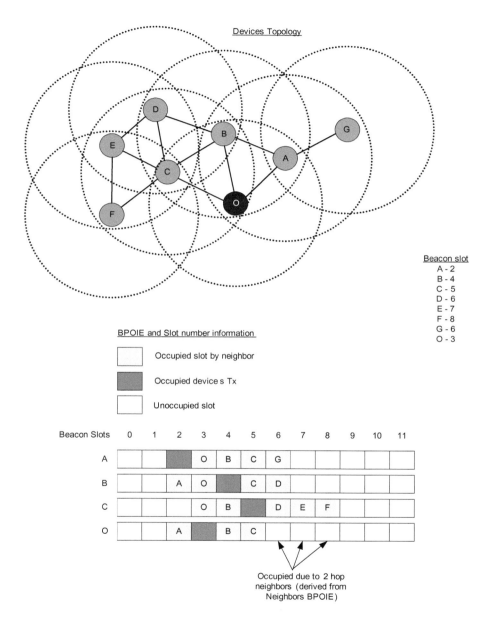

Fig. 5.20 BPO IE information usage.

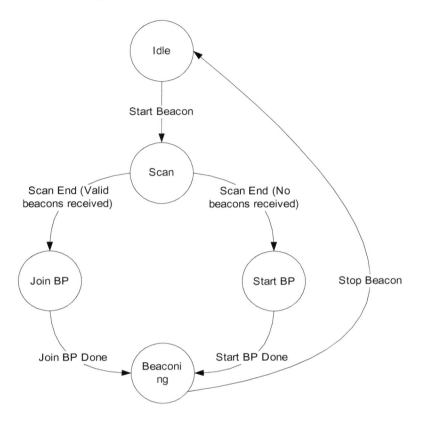

Fig. 5.21 Beacon state machine.

While joining the BP, device cannot select beacon slot higher than slot 47. This restriction is applied to limit the number of devices in 2 hop neighborhood. Too many devices in extended BG may result in insufficient time for data communication for each device.

5.5.5   Beacon Transmission

In normal situation, device transmits beacon once in a superframe in its selected beacon slot. During the beacon transmission it has to follow some rules that are described here.

Beacon Slot Selection: While joining a BP, device selects beacon slots. There are conditions on when device has to select new beacon slot during its operation. When the device selects new beacon slot due to slot collision or BP merge, it selects beacon slot up to a maximum BP length (up to slot number 95). If device cannot select beacon slot in this limit, then it stops the operation.

BP Length: As described in Figure 5.20, BP length of transmitted beacon includes own beacon slot and all highest unavailable beacon slots in 2 hop neighborhood. BP length can include some extra slots (up to 8) called extension slots to provide the room for new joining devices. Device keeps itself awake till BP length so it can receive the beacons from neighbors. Device can go in sleep after its BP length.

Signaling Slot Usage: It is possible that device's selected beacon slot is beyond one of the neighbor's BP length. If device transmits beacon in that selected beacon slot it may not be received by one of the neighbors. Solution to this problem is to send the beacon in one of the randomly chosen signaling slot. Each device is awake during the signaling slot, so the signaling beacon can be received by everybody. Device which has short BP length can use the information provided by signaling beacon to increase the length of BP. To avoid the collision of beacons in signaling slot, signaling beacon is maximally transmitted for 4 superframes continuously and then it is not transmitted at least 4 superframe again. The device transmitting beacon in the signaling slots also transmits the beacon in its chosen slot.

Skipping Beacon Transmission: UWB system is half duplex system so it can either receive or transmit. During transmission it cannot receive anything. If two devices present in network and both are transmitting beacons in same slot then both cannot listen to each other, In order to detect beacon collisions with such neighbor, the device skips its beacon transmission aperiodically, and listens for a potential beacon reception in its beacon slot.

5.5.6   Beacon Reception

Device listens for beacons during its BP length announced in last superframe. It indicates the beacon received in last superframe in its BPO IE. If it received the beacon in signaling slot in the previous superframe, it extends its BP length such that beacon slot indicated in signaling beacon is covered under BP length.

Information from received beacon is used in synchronization, beacon slot collision detection, beacon period contraction etc. Device can receive alien beacon during BP. Alien beacon is a beacon with BPST, which is not same as device's BPST.

5.5.7   Beacon Slot Collision

It is possible that more than one device in extended BG has same beacon slot due to the mobility and nature of wireless medium. This situation is beacon slot collision and it leads to loss of beacons. There are different ways to find potential beacon slot collision and as a corrective action, device selects new slot after highest unavailable slot in extended BG. Following are the conditions when a device detects beacon slot collision.

- If device's beacon slot is marked occupied in received beacon's BPO IE from neighbor, but the corresponding device address is neither device's device address nor broadcast address. It confirms that some other device in extended BG is beaconing in device's beacon slot.
- After skipping the beacon transmission in previous superframe, BPO IE of neighbor marks the slot occupied by some other device. It confirms that some other device in extended BG is beaconing in device's beacon slot.
- When skipping beacon transmission in current superframe, if beacon frame is received in device's beacon slot. It confirms that there is another neighbor device with beacon slot same as device's beacon slot.
- When signaling beacon received with beacon slot number set to device's beacon slot number then it confirms that there is another beacon in device's beacon slot.
- Some events which repeatedly occur, like receiving broadcast address for device's slot by neighbor's BPO IE, some PHY reception activity in device's slot while skipping the beacon, are indication of possible beacon collision. As corrective action, device can select new beacon slot, can skip the beacon transmission, or can transmit a signaling beacon.

5.5.8 Beacon Period Contraction

In active network situation, many new devices join the network which eventually increase BP length and similarly many devices also leave the network. New device always joins after highest unavailable slot and lower beacon slots become available after devices in lower slots leave. BP length keeps on increasing due to these events. BP contraction is a procedure which reduces the variable BP length and keeps the BP length optimal.

If device has at least one available slot in its extended BG, between its own beacon slot and signaling slot for the last 4 superframes then device considers itself movable. Device's ability to move to lower slot is indicated in its beacon frame by setting movable bit.

Device can decide to move its beacon to lower available beacon slot after at least 4 superframes after it has become movable. To actually move, the device has to be the device with highest beacon slot number which is movable in its extended BG. This rule eliminates the possibility of multiple devices moving to same lower slot and creating beacon collision. Figure 5.22 shows the situation where device is device with highest beacon slot and movable

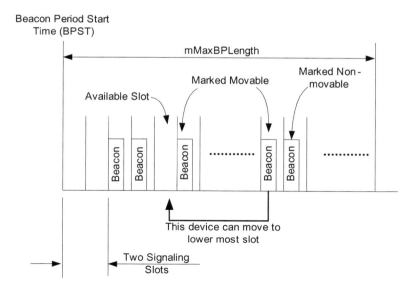

Fig. 5.22 Beacon period contraction (Move allowed).

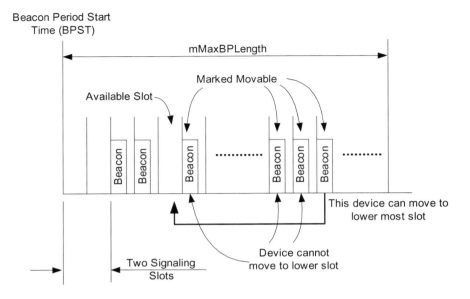

Fig. 5.23 Beacon period contraction (Move not allowed).

bit set, beacon can move to lower slot, while Figure 5.23 shows the situation where device cannot move the beacon to lower slot.

Device which includes Hibernation mode IE in its beacon are considered to be nonmovable during its hibernation period. The device which is involved in beacon slot collision or BP merge does not move to its lower slots in that transient network situation.

Another approach, though not in the MAC specification, could employ the announcement of intended move to lower available slot (thereby protecting that slot), before movement to the lower slot. This may provide multiple device movements at a time (Ref. [80]).

5.5.9 Synchronization using Beacons

Internal clock accuracy of different devices could be different, so superframe lengths for different devices are not same. To overcome this problem the device synchronizes with its slowest neighbor (largest superframe length) using beacons timing.

Devices in same BG maintain time reference, i.e. same BPST to communicate with each other. Beacon frame does not include any timestamp value, but timing is calculated using relative reception time of the beacon.

On reception of the neighbor beacon, device marks actual reception time actualRxTime. It also calculates the expected reception time expectedRxTime depending on slot number in the received beacon.

$$\text{expectedRxTime} = \text{Beacon slot number in received beacon}$$
$$\times \text{Duration of beacon slot } (85\,\mu s)$$

If expectedRxTime is less than actualRxTime then the neighbor is slower by (actualRxTime − expectedRxTime). Figure 5.24 shows this scenario. Similarly drift is calculated for each received beacons. Device delays its clock by maximum drift in each superframe, maximum up to $4\,\mu s$ in a superframe. If some beacons are missed during the superframe device can take historical information of earlier received beacons up to 3 superframes for synchronization purpose.

Another situation is when neighbor device is faster device, where expectedRxTime is more than actualRxTime. In this case device does not synchronize to the neighbor. However, for the neighbor, the device is relatively slower

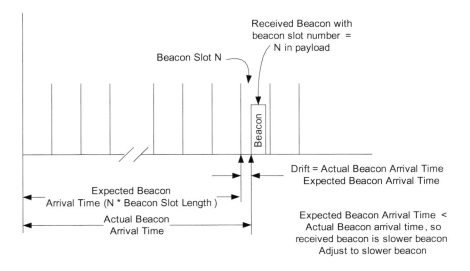

Fig. 5.24 Drift (Slower neighbor's beacon).

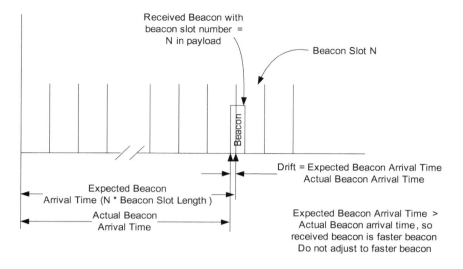

Fig. 5.25 Drift (Faster neighbor's beacon).

device so neighbor synchronizes to the device. Figure 5.25 shows faster neighbor scenario.

In case of bad channel, it is possible that drift increases too much then one of the devices becomes overlapping alien and BP merge procedure takes care of overcoming this un-synchronization.

Signaling beacon is not used for synchronization purpose, because signaling beacon does not have actual signaling slot number where beacon is transmitted, so expected time of reception cannot be found.

5.5.10 Beacon Period Merge

More than one different BPs come in vicinity of each other due to the change in environment and mobility. That creates overlapping superframes. It is not a suitable situation for proper data transfer and normal operation. Device may receive a beacon in DP or may continuously be subjected to the data or beacon loss.

If the BPST of the other device differs more than double of guard time (i.e. 24 μs) then that beacon is considered as alien beacon. Synchronization failure due to beacon loss can also cause the beacon of the fast device to appear as alien beacon.

5.5.10.1 Overlapping Alien BPs

If BPST of the device falls within the alien BP then it is overlapping alien BP, where the device relocates its beacon to alien BP in order to merge BPs and create single BP for proper operation and stops sending beacon in original BP. This rule forces only one of the devices to merge with another and avoids race condition of both merging each other. Figure 5.26 shows the overlapping BPs.

Device takes following steps to merge to overlapping BP.

- Device adjusts its BPST by forward difference of own BPST and alien BP's BPST as shown in Figure 5.26.
- Device selects new beacon slot as (old beacon slot number + 1 + number of highest occupied beacon slot in alien BP − number of signaling slots). It really places the old BP after new BP.

5.5.10.2 Nonoverlapping Alien BPs

Alien BP, which is not overlapping is nonoverlapping alien BP. When device detects nonoverlapping alien BP, it protects that alien BP with reservation type alien BP. No data is transmitted during the reservation type alien BP and the protection helps other device to receive the alien beacon.

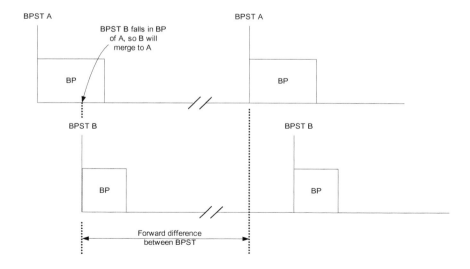

Fig. 5.26 Overlapping Alien BPs.

Nonoverlapping BP merge is announced through BP Switch IE. If any device in alien BP is transmitting BP Switch IE then alien BP is trying to merge so device does not try to merge but waits for alien BP to finish the merge operation. If BP Switch IE is not received from Alien BP then device tries to merge to alien BP.

It is also possible that alien BP is a transient situation due to mobility. Alien devices may move away and BP merge may not be required. So, it is suggested to wait for some time before device starts adding BP Switch IE. If alien BPST falls in device's superframe's first half then device need to finish BP merge in 128 superframes (as shown in Figure 5.27), else need to finish in 192 superframes (as shown in Figure 5.28). These rules are enforced to avoid the race condition in BP merge.

When device decides to start BP merge by adding BP Switch IE then it adds BP Switch IE with the following parameters.

- It sets BP Move Countdown to 9.
- BPST offset as forward difference of own BPST and alien BPST.
- Beacon slot offset as (number of highest occupied beacon slot number in alien BP + 1 − number of signaling slots).

BP Switch IE is updated in each superframe with current offset between BPs.

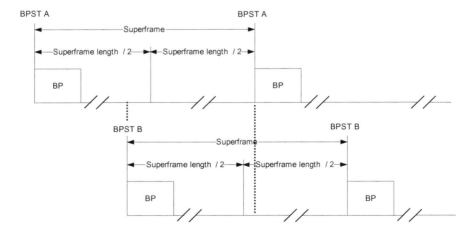

Fig. 5.27 Nonoverlapping Alien BPs (First half of superframe).

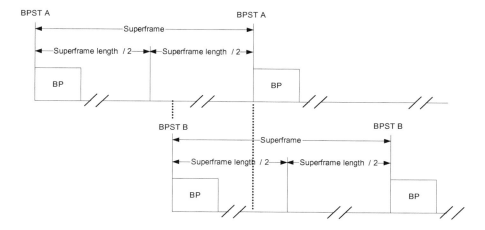

Fig. 5.28 Nonoverlapping Alien BPs (Second half of superframe).

Any time, during BP merge process, if it is found that alien BP is merging before the device merges to alien BP then device halts BP merge and waits for alien BP to merge.

Beacon Period merge is a synchronized procedure to move devices having same BPST to another BPST. It is also possible that all devices do not receive alien beacon but those devices receive the BP Switch IE from its neighbor. Devices follow that BP Switch IE and add same BP Switch IE in their beacons and perform the BP switch.

At the end of the superframe, when device includes BP Switch IE with move countdown field as zero, it adjusts its BPST according to BPST Offset field. It starts sending beacon in new BP in new beacon slot number as (old beacon slot number + Beacon slot offset).

Alternative to BP merge, device can always stop the operation in old BP and join as new device to alien BP.

5.6 Data Transfer Mechanisms

Medium Access Control (MAC) provides different methods for data transfer depending on applications and upper layer requirements. Data can be transferred through reservation based or contention based mechanism with different acknowledgment policies.

Data unit provided by upper layer to MAC is called MAC Service Data Unit (MSDU). MAC processes it, if required fragments it, and attach those with MAC header. The frame, which is sent to PHY, is called MAC Protocol Data Unit (MPDU). On the recipient side, MPDU is received by recipient's MAC. MPDU is processed here, if required reassembled, and it is called MSDU. MSDU is sent to upper layer. Similarly, data units of commands exchanged between MAC layers are called MAC Command Data Unit (MCDU). Commands are not sent to upper layer, it is internal to MAC control mechanism so generated and consumed at MAC layer itself.

Depending on implementation logic, channel conditions, historical information, the source device can fragment MSDU or MCDU. Size and number of fragments again depends on implementation. Fragments are required to be reassembled at reception side. For that purpose device maintains sequence control and fragment number. The device sets the fragment number field in the first fragment to zero and in subsequent fragments; fragment number is incremented by one. Fragment number is not incremented for retransmitted fragments. Sequence number for all fragments of same MSDU/MCDU is same.

The recipient device reassembles the MSDU/MCDU in the correct order before delivery to the upper layer. If the No-ACK policy is used and any fragment is missed then it is not possible to get it again, so the recipient device discards an MSDU/MCDU immediately if a fragment is missing. In other case device discards an MSDU/MCDU after implementation dependant timeout if a fragment is missing.

5.6.1 Reservation based Data Transfer

The data traffic with service interval requirement and isochronous stream requires dedicated channel time for data transfer. MAC provides reservation based mechanism for that. Reservation is a dedicated one-way channel time for data transfer. MAC defines the protocol reservation establishment between two or more devices. The protocol is called Distributed Reservation Protocol and used to reserve and maintain one or more MASs for unicast or multicast communication.

The device that initiates the reservation is called owner. Owner transmits the data traffic over the reservation. Device(s) that receives the traffic is called reservation target. If bi-directional stream is required then two different reservations, one in each direction, is required to be established. Reservation is uniquely identified by target device address, owner device address, reservation type, and stream index.

Different types of reservation are defined for different purposes, as described below:

Hard Reservation: The owner of the reservation can transmit data, command or control frames to the target of the reservation during hard reservation. Unused hard reservation can be released to other devices for use.

Soft Reservation: During soft reservation, any devices can transmit data following contention based PCA rules, but the reservation owner can transmit the data, control or command frame to the target with the highest priority and without performing back-off. Owner of the reservation start transmission at the start of the reservation block. Details of PCA are described in the following subsections.

Alien BP Reservation: Alien BP reservation is used for protecting alien BP. It is the highest priority reservation and requires no negotiation. No frames are transmitted in alien BP reservation. Device can only transmit beacon during alien BP reservation for the purpose of merging to alien BP.

Private Reservation: Private reservation is a special kind of reservation defined for upper layers and applications. Channel access method for the private reservation is defined by the upper layer which is using the private reservation. Standard MAC frame format are used for the data communication during this reservation. Devices other than owner and target cannot access the medium during private reservation. Release of the reservation is similar to hard reservation, after release it can be used by any device.

PCA Reservation: Any device can access the medium during PCA reservation using Prioritized Contention Access (PCA) method. This reservation is one kind of anti-reservation, where devices can access the medium using PCA but cannot make any other type of reservations. It confirms the PCA time during the superframe. PCA reservation does not require any negotiation. It can be created on any unreserved area or overlapping to any other PCA reservation of the neighbor.

5.6.1.1 Hidden Node for the Reservation

Data transmission in decentralized topology suffers from hidden node problem. Due to hidden reservation, it is possible that reservation negotiation may fail or conflict of the reservation occurs.

Figure 5.29 illustrates an example of hidden node problem. Three devices A, B, C, and D are shown with their radio range. Device A can communicate with B and not directly with C, but B can communicate with A and C. If there is a communication between A and B, it can be interfered by communication between C and D.

Reservations have to be in such a way that there are no overlapping reservations between A–B and C–D. Due to simultaneous negotiation or mobility of the devices if these reservations exist at the same time then reservation conflict resolution mechanism resolves that conflict. Negotiation and conflict resolution mechanism takes care of hidden reservation, which is described in the following subsections.

Distributed Reservation Protocol Availability IE can help in reservation negotiation with providing information of extended beacon group's reservation. DRP Availability IE provides information about the available MASs (unreserved by itself, unreserved by neighbors and not part of BP) around the device. The combination of information from received DRP Availability IEs and received DRP IEs allows an owner to determine appropriate MASs for a

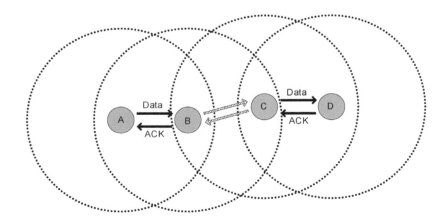

Fig. 5.29 Hidden node problem.

new reservation. Owner only sends the request for MASs which are available for itself and available to potential target of the reservation. This reduces the possibility of the rejection of the reservation due to conflict at target side.

The traffic without any acknowledgment can provide one more level of spatial reuse. Multicast traffic uses No-ACK policy. Though not specified by the standard, enhanced usage of DRP Availability IE can help to achieve the spatial reuse for data transmission for the traffic without any acknowledgment (Ref. [76]).

5.6.1.2 Reservation Negotiation

Reservation negotiation can be performed through beacons or command frame. Beacon is transmitted only once in a superframe, while command can be multiple in a superframe, that is why negotiation through commands is comparatively quick. Negotiation through beacon takes at least two superframes (Reservation Request and Reservation Response). Commands are generally transmitted through PCA. It can be transmitted through reservation also, but it requires already established reservation between owner and target in both directions.

New reservation is initiated by the reservation owner by transmitting a DRP IE with selected available MASs (using 2 hop information provided by DRP Availability IE), target/owner device address is set to device address of the target of the reservation, reservation status to 0, reason code to accepted. Stream index is set to unique value which is not used for same target.

On reception of this DRP IE, target device can accept or reject the new reservation request. Target can reject the reservation request due to different reasons, like unavailability of the requested MASs, reservation policy violation (described later in the chapter), internal resource availability etc. DRP Availability IE is also included by target on rejection.

If the unicast reservation is accepted by the target, then target adds DRP IE with target/owner field set as device address of the owner, reservation status set to 1, reason code set to Accepted and rest fields set same as request. Figure 5.30 shows unicast reservation negotiation procedure in form of sequence diagram and reservation establishment steps.

If reservation request is for the multicast target then it is possible that subset of devices accept the reservation and rest may reject it or subset of

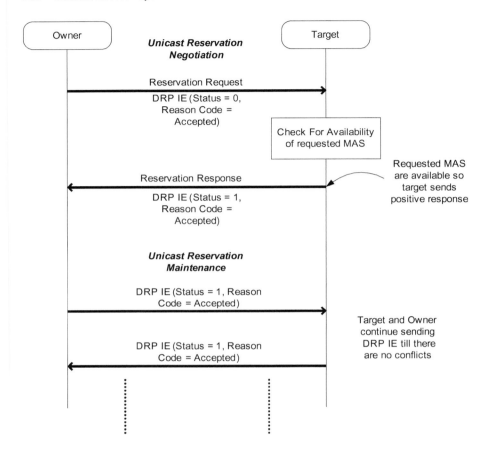

Fig. 5.30 Unicast reservation negotiation.

requested MASs are accepted. On acceptance of multicast reservation, device adds DRP IE similar to unicast with status 1. Multicast reservation target always includes DRP Availability IE with multicast reservation response to facilitate reservation negotiation. Multicast reservation owner has to take care of more number of targets and wider 2 hop neighborhood, so DRP Availability IE is very helpful to select the MASs without any conflict for the target. Figure 5.31 shows reservation negotiation procedure for multicast reservation.

Multicast reservation owner can set the reservation status to 1 regardless of the response from the targets, because owner tries to serve as many as possible devices through multicast reservation. Different target devices may

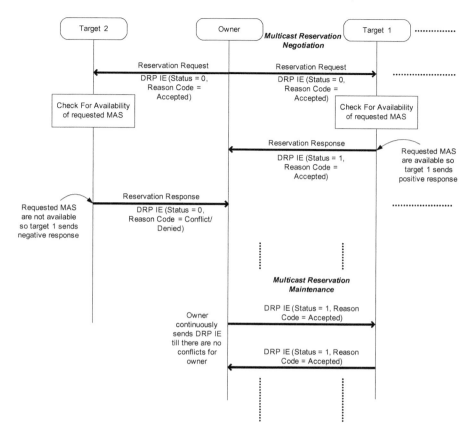

Fig. 5.31 Multicast reservation negotiation.

have rejected or accepted reservation with different accepted MASs. Owner internally maintains the different accepted MASs from different targets to schedule the traffic to all devices. Owner may have to retransmit some traffic in case common reservation is not available for all multicast targets.

5.6.1.3 Reservation Maintenance

Once reservation negotiation is completed and reservation is established (status set to 1, reason code as Accepted), both the owner and target(s) includes DRP IE in its beacon in each subsequent superframes. DRP IE is included in beacon till reservation termination. If reservation is modified then same DRP

IE is modified. Owner and target device always check for reservation conflict during reservation establishment and maintenance. If conflict is found then conflict is resolved according to conflict resolution rules.

5.6.1.4 Reservation Modification

Device may require increasing, decreasing, or moving the MASs of the reservation depending on application requirement, policies or change in scenario. If reservation owner requires more MASs or move the location of the reservation for the stream, then it can negotiate the additional MASs using negotiation rules with same target/owner device address, stream index, and reservation type. After successful negotiation, owner combines the DRP IEs with reason code set to Modified. Target of the reservation also changes the DRP IE according to owner's DRP IE. Figure 5.32 illustrates the reservation modification procedure for adding extra MASs to reservation in form of sequence diagram.

Reservation owner is not required to change the reservation status or not required to renegotiation reservation if it removes some MASs from its established reservation. Owner removes MASs from its DRP IE and set reason code to modified until the reservation target changes its DRP IE to match owner's DRP IE. Figure 5.33 shows the modification procedure.

If modification is initiated by the target, then it is not required to change reservation status, while removing some MASs from its reservation. It is enough to set reason code as modified. If the owner is unicast then owner removes the MASs as per target and match the DRP IE with reservation status set as 1 and reason code to Accepted or owner can terminate the reservation. In case of multicast, reservation owner may not change the DRP IE but considers that target has changed the DRP IE for sending data.

5.6.1.5 Multicast Reservation Joining

If a device wants to receive some multicast stream then it can join already established multicast reservation and become a target. Target can join the multicast reservation without going through negotiation if reservation does not conflict with any other reservations in the neighborhood. Target just have to add a DRP IE with reservation status 1 and reason code as Accepted for joining multicast reservation.

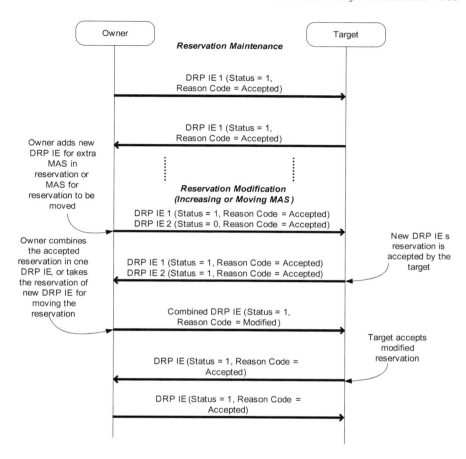

Fig. 5.32 Reservation modification (Adding additional MASs).

5.6.1.6 Reservation Conflict Detection and Resolution

Multiple simultaneous reservation negotiation or mobility of the devices can result in conflict of reservations.

Alien BP reservation does not conflict with any other reservation. Reservation conflicting with alien BP reservation is not used for data communication and set with reason code conflict. PCA reservation also does not conflict with any other PCA reservation.

If already established reservation (status 1) conflicts with reservation under negotiation (status 0), then established reservation wins the conflict and used

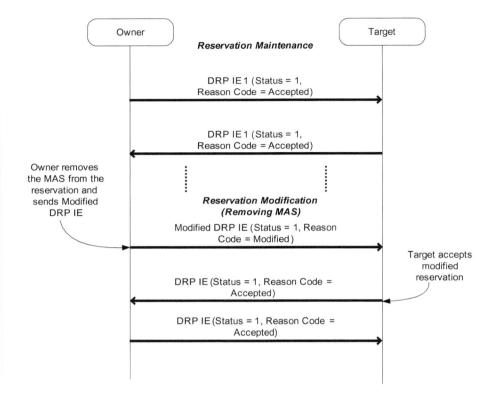

Fig. 5.33 Reservation modification (Removing MASs).

for data communication. Loosing reservation negotiation is not continued and can be renegotiated.

If the conflicting reservations have same value for status bit then one of that wins and set the status 1. If conflicting devices' tie-breakers are same, then device with lower beacon slot wins the reservation conflict. If conflicting devices' tie-breakers are different then device with higher beacon slot wins the reservation conflict.

The loosing device sets reservation status 0 and reason code to Conflict for the reservation. Whenever reservation target sets reason code to Conflict in any DRP IE in its beacon, it includes a DRP Availability IE to facilitate the selection of new MASs for resolution of the conflict. Whenever a device releases a reservation or part of reservation due to conflict, it invokes back-off procedure (in number of superframes) prior to requesting additional MASs in any of its reservations.

5.6.1.7 Reservation Termination

Owner can terminate the reservation any time by removing DRP IE from its beacon. In that case target of the reservation considers the reservation being terminated and removes the corresponding DRP IE.

Reservation target actually cannot terminate the reservation by itself, because it does not know about the requirement of application at owner side. But, it can indicate the nonacceptance of the reservation by setting appropriate reason code and reservation status to 0. Owner either terminates the reservation and sets its own reservation status to 0 or renegotiates the reservation.

In another case, if reservation owner or target do not receive the beacon and the reservation is being unused for more than 3 superframes, then reservation is considered to be terminated and DRP IE is removed.

5.6.1.8 Reservation Policies

Superframe is an open medium; device that comes first gets the first chance to make reservation. In that case it is possible that first device hogs the channel with complete superframe reservation and stops any other device to operate in the channel. MAC defines reservation policies to provide fair access of the medium to maximum possible devices. Policies also take care of service interval and latency requirements of different streams by positioning the reservations in such a way that it has possibility to accommodate new reservations and also support the requirements of existing reservations.

Figure 5.15 shows the zone structure of the superframe. Reservations can have row components and column components. Row component includes equal number of MASs in every zones of the superframe, exceptionally excluding zone zero. The reservation component which is not row component is column component. Figure 5.34 shows row and column reservation components.

Reservation either can be safe or unsafe depending on policies. Unsafe reservation is relinquished if requested by any neighbor, but safe reservation is not required to be relinquished from any neighbor's request. Safe and unsafe

Beacon Zone

Allocation Zone

	0	1	2	3	4	5	6	7	8	9	10	11	12	13	14	15
0	0	16	32	48	64	80	96	112	128	144	160	176	192	208	224	240
1	1	17	33	49	65	81	97	113	129	145	161	177	193	209	225	241
2	2	18	34	50	66	82	98	114	130	146	162	178	194	210	226	242
3	3	19	35	51	67	83	99	115	131	147	163	179	195	211	227	243
4	4	20	36	52	68	84	100	116	132	148	164	180	196	212	228	244
5	5	21	37	53	69	85	101	117	133	149	165	181	197	213	229	245
6	6	22	38	54	70	86	102	118	134	150	166	182	198	214	230	246
7	7	23	39	55	71	87	103	119	135	151	167	183	199	215	231	247
8	8	24	40	56	72	88	104	120	136	152	168	184	200	216	232	248
9	9	25	41	57	73	89	105	121	137	153	169	185	201	217	233	249
10	10	26	42	58	74	90	106	122	138	154	170	186	202	218	234	250
11	11	27	43	59	75	91	107	123	139	155	171	187	203	219	235	251
12	12	28	44	60	76	92	108	124	140	156	172	188	204	220	236	252
13	13	29	45	61	77	92	109	125	141	157	173	189	205	221	237	253
14	14	30	46	62	78	94	110	126	142	158	174	190	206	222	238	254
15	15	31	47	63	79	95	111	127	143	159	175	191	207	223	239	255

143	Row reservation component
134	Column reservation component

Fig. 5.34 Row and column reservation components.

are indicated in DRP IE of the reservation. Reservation is marked as unsafe according to following rules, rest of the reservation are safe reservation.

- If device (regardless of owner or target) reserve more than 112 MASs then that reservation is "over reservation" by single device. Extra MASs more than 112 are marked unsafe.
- Figure 5.35 shows the column components of the reservation starting at different MASs. Dark color indicates that component as safe. Reservation component with higher MAS number in a zone is unsafe. In other words, a device (regardless of owner or target) cannot allocate more than Y (Column in dark color) consecutive MASs

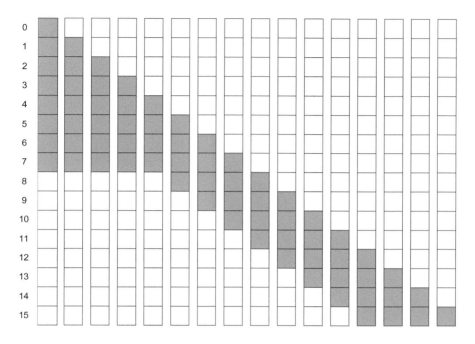

Fig. 5.35 Column reservation limits.

as safe in the same zone, where Y depends on lowest reserved MAS number in the particular zone as shown in Figure 5.35.

From the above rules, it is possible that a reservation which is safe for owner could be unsafe for target and vice-versa.

Few other points related to policy are as follows:

- Zone zero (beacon zone) can be used only for row reservation.
- For alien BP reservations, no policies apply and alien BP reservation MASs are not even counted for policy of 112 MASs.
- Device should include PCA reservation similar (overlapping) to neighbor if it wants to use that period for frame transfer using PCA.
- For safe row reservation, owner makes reservation in lowest available row. If lower rows (lower rows in zone structure of superframe) become available then owner moves its row reservation to available lower rows in order to make the room for possible new reservations.

Table 5.5 Zone index and service interval.

Zone index	Service intervals (in Unit of zones)
8	16
4, 12	8
2, 6, 10, 14	4
1,3,5,7,9,11,13,15	2

- Depending on service interval requirement and availability of MASs in the superframe, owner selects the safe reservation pattern. Table 5.5 shows the zone index and service interval. On selection of reservation block(s) in particular zone, service interval from that column can be achieved as shown in the following table.
- If there are multiple possibilities to achieve the service interval then zone index is selected from upper rows in Table 5.5. That means owner follows the sequence [{8}, {4 or 12}, {2, 6, 10, or 14}, {1, 3, 5, 7, 9, 11, 13, or 15}] for selection. Owner tries to make the column reservation in upper 8 MASs of the each zone in MAS map table in order to make it safe reservation.
- All safe column reservations are made as much higher MAS (higher in zone structure of superframe) as possible in order to make the room for possible new reservations, If some higher MASs become available then reservation owner moves its reservation to higher MASs.
- Unsafe reservation can be made anywhere.

5.6.1.9   Reservation Relinquish

Device requires some number of MASs for a data stream. If enough MASs are not available then the device can send relinquish request to the unsafe reservations. Relinquish request IE is sent to the device (could be either owner or target of the reservation) with unsafe reservation. Allocation field of the IE includes MASs requested to be relinquished.

If device receiving relinquish request is the target of the reservation then it forwards the request to owner of the reservation with DRP Availability IE. Target does not have information about the requirements of the reservation at owner side, so relinquish request is forwarded to owner.

On reception of Relinquish Request IE by owner of unsafe reservation, the owner either removes the requested MASs from its reservation or modifies its reservation such that all its reservation become safe.

If the device is sending relinquish request for some unsafe reservation of neighbor, then the device cannot make any unsafe reservation immediately. Similarly, a device receiving relinquish request for some unsafe reservation, cannot make new unsafe reservation immediately, after relinquish.

5.6.2 Priority Based Data Transfer

One of the basic access mechanisms for UWB MAC is CSMA/CA to serve the bursty and priority based traffic with best effort. It is very similar to contention access method defined by IEEE 802.11 MAC. Contention based data transfer mechanism defined by UWB MAC is called Prioritized Contention Access (PCA). It is a variant of collision avoidance by listening before transmitting and binary exponential back-off. PCA is a Prioritized Contention Access method where multiple devices contend for the medium and winner uses the medium for data transfer. Four different access categories also internally contend for accessing the medium. PCA supports prioritized QoS. Traffic to be transmitted is mapped to one of these access categories based on its priority.

Device can choose to transmit in a PCA period, i.e. unreserved period other than BP, PCA reservation, and soft reservation. Device performs CCA using preamble sensing. If something is received (preamble received) during CCA, it means medium is busy and device does not transmit anything on air. It is possible that preamble part of the frame is not received during CCA then CCA falsely detects the clear channel, which can lead to data collision. In case of collision, device invokes binary exponential back-off algorithm and retry for the transmission with incrementing the retry counter.

If device detects medium available through CCA, then it waits for medium to become idle. Waiting period is called Arbitration Inter-Frame Space (AIFS) shown in Figure 5.36. AIFS for access category with lowest priority is longest and for highest priority it is shortest.

After AIFS time is over, device has to perform the back-off procedure for the access category before transmitting the frame. Back-off procedure is explained in the following subsections.

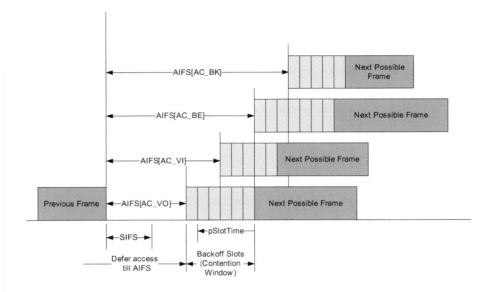

Fig. 5.36 AIFS.

When device gets the opportunity to transmit the frame on air, it is called transmission opportunity. Transmission opportunity remains with same device and access category for transmission opportunity limits time duration defined in MAC standard. Multiple frames can be transmitted in one transmission opportunity. Transmission opportunity is considered to be over when SIFS + Guard time before PCA time ends.

5.6.2.1 Back-off Procedure

Device maintains the back-off counter for each access category. Back-off procedure is as follows:

- Device starts new back-off when a new frame arrives in its internal buffers for transmission after buffer was empty. For that Access Category (AC), device selects contention window, CW[AC] as mCWmin[AC], and selects random back-off counter between [0, CW[AC]].

- Device has frame in buffer to transmit but there is not enough time in transmission opportunity then device can use same CW[AC] to select back-off counter between [0, CW[AC]].
- If internal or external collision of frame happens then device selects new back-off counter. For that, it selects new CW[AC] smaller from $2 \times$ CW[AC] $+ 1$ or mCWmax[AC] and selects back-off counter between [0, CW[AC]].
- If device does not receive expected CTS or ACK then also it selects new CW[AC] smaller from $2 \times$ CW[AC] $+ 1$ or mCWmax[AC] and selects back-off counter between [0, CW[AC]].
- After transaction is over, device sets the CW[AC] again to mCWmin.

After waiting for AIFS[AC] time after medium was idle, device starts decrementing back-off counter as shown in Figure 5.36. Decrementing back-off counter means waiting time duration of a slot of $9\,\mu s$. During this time device has to perform CCA on start of each slot time. If medium is idle then back-off is continued till it reaches zero. If medium activity is found then back-off counter is frozen and continued when medium is available again. If back-off counter of more than one access category becomes zero in the device then AC with higher priority gets transmission opportunity and back-off counter for other access category is frozen.

5.6.2.2   Virtual Carrier Sensing

Device uses Request to Send (RTS) and Clear to Send (CTS) control frames before transmitting data frames to reduce the possible collision due to hidden node problem. RTS CTS frames prevent neighbors of transmitter and receiver to access the medium during data transfer between transmitter and receiver.

Device transmits RTS frame on start of transmission opportunity. Duration value of RTS includes the time required for frame transactions, i.e. interframe spaces, CTS frame, Data frame, and ACK frame. Device receiving RTS frame other than intended recipient sets Network Allocation Vector (NAV) and does not access the medium till NAV is zero (expired).

Device receiving RTS, responds with CTS frame, if its NAV is zero. CTS frame duration field includes time required for frame transaction, i.e. less of

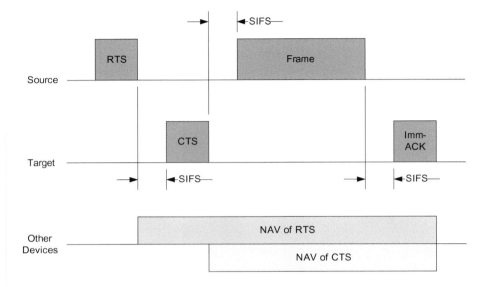

Fig. 5.37 RTC CTS usage.

CTS frame and SIFS from duration of RTS frame. If NAV is not zero on reception of RTS frame then CTS is not sent and frame transaction cannot be processed. All devices receiving CTS other than intended target of CTS set NAV using duration value of CTS. During the NAV those devices do not access the medium.

On reception of expected CTS transmitter device can send data frame. If expected CTS is not received then device can retransmit RTS after end of expected CTS end time plus SIFS, after back-off.

This procedure is called virtual carrier sensing. Figure 5.37 shows usage of RTS CTS for virtual carrier sensing.

5.6.2.3 Data Transfer using PCA

Device can transmit and receive multiple data or command frames, acknowledgments, RTS and CTS during transmission opportunity. Device even can retry the frames before transmission opportunity limit is over. If device cannot retry the frame in same transmission opportunity, due to transmission opportunity limit is over or PCA period is over, then it can transmit in next transmission opportunity.

Source device also considers the target's availability of receiving PCA traffic in selected MASs, provided through PCA Availability IE. TIM IE is added in beacon by source of data traffic to indicate availability of traffic for a particular target device. TIM IE and PCA Availability IE are used in conjunction with each other. If TIM IE required bit in the recipient PCA Availability IE is not set then recipient is always available during announced PCA MASs. If TIM IE required bit in recipient PCA Availability IE is set then transmitter device has to send TIM IE for buffered traffic otherwise recipient does not remain awake during announced MASs also.

5.7 Summary

This chapter covers key requirements for UWB MAC system, different options considered for MAC system, and impact of those features on overall system. Different type of MAC features are discussed in this chapter for its usage considering UWB system.

This chapter also explains the medium access and control protocol of UWB MAC system based on ECMA-368. It explains the decentralized network topology without any coordinator with the help of distributed beacon protocol. Also, different methods for data transfer (reservation based and contention based) for different type of traffic were discussed.

This whole system together creates a reusable MAC for multiple simultaneously active upper layers so it is called UWB common radio platform.

6

Advanced MAC Features

6.1 Introduction

Functions like beaconing and data transfers are the base of MAC protocol and required for interoperation and data communication. There are many other advanced MAC features defined to meet the UWB system requirements in regards to mobility, power save, security etc. They also provide optimized performance with different acknowledgment policies, aggregation, transmit power control, and enable range measurement support. Following subsections describe these features for UWB MAC.

6.2 Acknowledgment Policies

MAC supports three different acknowledgment policies for the verification of the delivery of a frame. Three acknowledgment policies are No-ACK, Immediate-ACK (Imm-ACK), and Block-ACK (B-ACK). Depending on the application requirements, one of these acknowledgment policies can be chosen for the data stream.

No acknowledgment can be used for multicast and broadcast frames, because acknowledgment frames from multiple targets can collide and transmitter may not receive any acknowledgment frames.

6.2.1 No-ACK

No-ACK acknowledgment policy is used with the traffic where assured reception of the frame is not required or late arrival of the retransmitted frame is not useful. For video or audio traffic, it is safe to miss some frames rather than frames reaching late after retry. Sometimes, upper layer of MAC is handling required acknowledgment, so No-ACK can be used.

Device transmitting frame with No-ACK policy always considers that frame is received by other side successfully. No retry is done for the frame, so no frames are buffered for retry at source.

Device receiving frame with No-ACK policy does not transmit any acknowledgment and sends completely formed MSDU to upper layer without any reordering. If any of the fragments is missed in MSDU then remaining fragments of that MSDU are discarded, because without retry there is no way to obtain the missed fragment.

6.2.2 Immediate-ACK (Imm-ACK)

Imm-ACK acknowledgment policy can be used with traffic where assured reception of the frame is required. Imm-ACK is used with unicast frames only.

Device transmitting frame with Imm-ACK policy is buffered for possible retransmission. If device does not receive Imm-ACK frame after transmission of the frame then device retransmits the frame. Number of retransmission (retry of the frame) is implementation dependent.

Device receiving frame with Imm-ACK policy transmits Imm-ACK frame in response to received frame after SIFS duration.

6.2.3 Block-ACK (B-ACK)

Block-ACK or B-ACK is used when assured reception of traffic is required but frequency of too many acknowledgment frames (Imm-ACK) is an overhead. B-ACK mechanism transmits only one acknowledgment frame in response to reception of multiple frames. Depending on received B-ACK frame, the source device retransmits required frames from buffered frames. Different B-ACK sessions can be established for different DRP or user priority simultaneously. It requires a dedicated sequence number for each B-ACK session. B-ACK is a complex feature so all devices may not support it. Capability of device to support B-ACK is indicated in its MAC Capabilities IE.

Source device starts B-ACK session by transmitting frame with B-ACK Request acknowledgment policy. Recipient device can either accept or reject the usage of B-ACK. Rejection could be because of not supporting B-ACK or unavailability of resources at that time. Recipient transmits B-ACK frame with no payload for rejection. Otherwise recipient transmits B-ACK frame with allowed maximum size of the frame for next B-ACK sequence with acknowledgment of the received frame in the acknowledgment window. If source does not receive any B-ACK in response then it can retry the frame.

If recipient device does not have buffers for new frames and wants to continue for B-ACK mechanism then it transmits B-ACK frame with frame count or buffer size set to zero. This indicates busyness or temporary unavailability of the resources at the recipient device. In response to that, the source can either transmit same frame again or transmit the dummy frame (same frame with same sequence number with zero length payload) with sequence number same as earlier.

During the normal operation of B-ACK, recipient device transmits B-ACK with available frame count and buffer size for the B-ACK session. Source device considers all frame received before the sequence number indicated in the B-ACK frame and it can remove from its internal buffers (buffered for possible retransmission). Source also considers all frames received which are indicated as one in the bitmap of acknowledgment window. Remaining frames in the internal buffers of source are to be retransmitted according to increasing sequence and fragment number.

After first B-ACK frame from recipient with frame count and buffer size, source transmits all frames with acknowledgment policy set to B-ACK and last frame with acknowledgment policy set to B-ACK Request. A frame with B-ACK request is transmitted before frame count or buffer size reaches to target's frame count or buffer size.

Recipient transmits B-ACK in response to B-ACK request with current frame count and buffer size available for the B-ACK sequence, bitmap of received frames (set to one) and missed frame (set to zero) in acknowledgment window, and start of the window by sequence number. Recipient device reorders and reassembles the received frames and sends MSDU to upper layer. It is also possible that recipient implements timeout for the frame hold in buffer waiting for other frames, so after timeout those buffers can be used for new frames and acknowledgment window can advance.

Source can terminate the B-ACK usage by using any other acknowledgment policy other than B-ACK or B-ACK request. Recipient can terminate the B-ACK usage by transmitting B-ACK frame with no payload.

6.3 Power Save

Power save is very important for any battery operated wireless device. When device is idle and not communicating with any other device then it can save power by switching off the radio and moving system to nonactive state.

MAC provides techniques for power saving. In order to coordinate power save with neighbors, the device transmits Hibernation mode and Hibernation Anchor IE. Device either go in sleep for idle time during superframe called intra-superframe power save or go in sleep for multiple superframes called inter-superframe power save.

6.3.1 Intra-superframe Power Save

Device has to be awake during any transmission or while receiving something. During the idle time (no transmission or reception) device can sleep. MAC protocol ensures about reception time and transmission time through reservation based mechanism, TIM IE and PCA Availability IE. Device can easily find the time duration when it can sleep for saving power consumption.

Following are some of the events during which device is awake during the superframe for its operation, rest of the time it can sleep.

- Device is awake during BP for transmission and reception of beacons. It wakes up before guard time to take care of clock drifts.
- If device is owner of the reservation then it wakes up for DRP traffic transmission before start of DRP block. It can go to the sleep after DRP block ends or transmission is over.
- If device is target of the reservation then it wakes up for DRP traffic reception before guard time before start of DRP block. Device can go to the sleep after DRP block ends or reception is over.
- Device wakes up before intended PCA time, if it has traffic pending to be transmitted through PCA. It can go to sleep after the transmission is over.
- Device wakes up for PCA traffic reception before guard time before start of PCA period, if device is expecting reception of traffic via PCA. If all data is received from the source then device can go in sleep.
- Device is awake for transmission and reception of acknowledgments.

6.3.2 Inter-superframe Power Save

Inter-superframe Power Save is also called hibernation mode. Device announces its intention to go to hibernation before going to hibernation mode.

During hibernation mode device does not transmit beacons or any other data traffic. Device also terminates all its reservations before going to hibernation. While the device is in hibernation, neighbors do not schedule any traffic or communication with hibernating device till hibernating device wakes up again. Neighbors of the hibernating device can buffer the traffic for hibernating device till hibernation is over. For multicast traffic it is suggested not to wait for hibernating devices.

Device includes Hibernation Mode IE in its beacon for multiple superframes before going to hibernation mode. Hibernation duration field is set to the number of superframes device is intended to be in hibernation mode. Countdown field indicates when device will be in hibernation mode. Countdown is reduced by one on each superframe. When it reaches zero, the device goes to hibernation from the next superframe.

Neighbor of the hibernating device protects hibernating device's beacon slot, for hibernation duration, by marking the beacon slot of hibernating device as occupied in its BPO IE. If neighbor receives beacon from any other device in hibernating device's beacon slot, still it marks slot as occupied by the hibernating device.

Device wakes up at least one superframe before hibernation duration is over in order to resynchronize itself with other devices beacons. If the beacon slot number of the device before hibernation is still unoccupied, then device uses that beacon slot and continue operation, otherwise it rejoins the network in the same manner a new device does. Device can come out of hibernation before hibernation duration is over.

If the neighbor does not receive beacon of the hibernating device after hibernation duration over, then it does not mark it occupied in BPO IE. But, internally it considers that slot occupied for 128 additional superframes and protects it.

6.3.3 Hibernation Anchor

A device that can provide the information about neighboring hibernating devices is called hibernation anchor. This information can be used by newly joined devices or devices those just woke up, to get the information about wakeup time of the hibernating devices. The device which can perform as hibernation anchor includes the information about its capability of acting as hibernation anchor in MAC Capabilities IE.

Device acting as hibernation anchor includes the hibernation anchor IE on reception of Hibernation Mode IEs from neighbors with hibernation countdown set to zero. Hibernation duration field in Hibernation Mode IE is set as wakeup in Hibernation Anchor IE with respective device address of hibernating device. Wakeup countdown is decreased by one on each superframe. When wakeup countdown field reaches to zero, the hibernating device is expected to wakeup in next superframe and resumes beaconing. Hibernation anchor removes the entry of the hibernating device whose wakeup countdown reached to zero, from Hibernation Anchor IE. If hibernating device wakes up earlier than its announced hibernation duration then also hibernation anchor removes its entry from the Hibernation Anchor IE.

6.4   Reservation Release (UDA/UDR Procedure)

If traffic arrival rate is low from upper layer, it is possible that some time of reservation is unused in the superframe. This unused time period in hard or private reservation block can be released. Owner can release the reservation which can be used by its neighbors by sending Unused DRP Reservation Announcement (UDA) frame. Target can release the reservation to its neighbors by sending Unused DRP Reservation Response (UDR) frame in response to UDA frame. This exercise should only be taken up if remaining time in the reservation block is sufficient enough for the exchange of UDA and UDR frames, and after that big enough that it can be really used by devices.

For unicast reservation release, owner transmits the UDA frame. UDA frame includes the target device which is supposed to send the UDR frame in response as target of the device. For multicast reservation release, UDA frame includes the list of device address of all the targets in multicast group, those devices are expected to send the UDR frames in the response.

The protocol takes care that UDR from multiple devices do not collide. Target device checks, if its device address is present in the UDA frame's list.

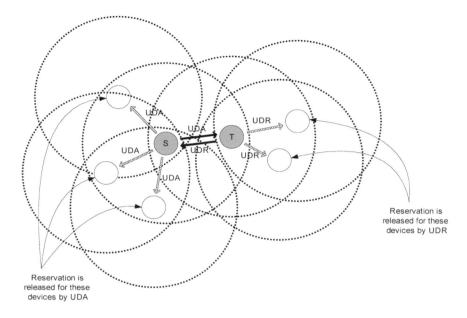

Fig. 6.1 Reservation release.

If it is present then it sends UDR frame with following delay.

Time to send UDR = SIFS time + slot time

+ (Position in list in UDA)

× (UDR control frame duration + SIFS time).

Reservation is considered to be available for current superframe, to those devices that listen to UDA or UDR frame, after last UDR frame is transmitted. In other words, device which receives UDA or UDR frame can consider medium available for PCA after duration value of UDA or UDR is elapsed. Duration of UDA includes UDA and all UDRs. Duration of UDR includes itself and other remaining UDR according to received UDA. Figure 6.1 shows how UDA and UDR can release the unused reservation for neighbors of owner and target.

6.5 Security Considerations

Wireless network is different from security point of view compared to wired network, because of not having wire or closed end to end medium. Wireless is an open medium where any compatible receiver can capture the traffic on the channel. To overcome the security weakness of wireless medium, MAC defines security mechanism by data encryption, message integrity, and replay attack protection.

Following are some of the methods defined by MAC which are used to provide the above mentioned security services:

- MAC defines three different security modes for different kind of security requirements of the applications and upper layers using MAC.
- It defines 4-way handshake mechanism between pair of devices to generate Pair-wise Temporal Keys (PTKs) based on pre-shared master keys. PTK is used for protection of unicast traffic between the pair of the devices.
- It also defines the way to distribute Group Temporal Keys (GTKs) using PTKs. GTK is used for the protection of multicast or broadcast traffic among group of devices.
- Secure frame counter and replay counter is used for message freshness which avoids replay attacks.
- It defines usage of Message Integrity Code (MIC) usage using Counter Mode Encryption and Cipher Block Chaining Message Authentication Code (CCM) for message integrity.
- Standard employs 128-bit symmetric temporal keys based on AES-128 for data payload encryption.

MAC does not provide any standard mechanism to address denial of service attacks, because denial of service at MAC level could be through channel hogging and medium disturbance.

6.5.1 Security Modes

Security mode of the device indicates whether the device is required to create secure relation for data communication with other device or not. The device has to perform 4-way handshake and generate PTK for the secure relationship. Different security modes are explained here.

Security Mode 0 (*Unsecured*)

- Security mode 0 is unsecured mode.
- It does not use any secured frame during communication.

Security Mode 1 (*Loosely Secure*)

- Security mode 1 device can create secure relationship with another security mode 1 device, but not with any other security mode devices.
- It uses secured frames for communication with security mode 1 device with which secure relationship is already established.
- It uses unsecured frames for communication with security mode 1 device with which secure relationship is not established or security mode 0 device.
- It cannot communicate with security mode 2 devices.

Security Mode 2 (*Fully Secure*)

- Security mode 2 device only communicates with another security mode 2 device with secure relationship.
- It only uses secured frame for data communication with security mode 2 device with secure relationship.

6.5.2 Temporal Keys

Different kind of temporal keys are used for secure relationship and secure communication.

Pairwise Temporal Key (PTK): Two devices generate secure relationship through 4-way handshake and create the PTK on both the sides using shared master key and two random numbers known as "Nonce." Pair of devices is considered to have secure relationship till they both have PTKs.

Group Temporal Key (GTK): GTK is distributed to a group of the devices through a message secured by PTK. GTK is used by the group of devices for secure communication.

Master Key Identifier (MKID): Device can have multiple pre-shared master keys and those master keys are identified by MKID. Device can share those MKID through beacons or can be probed whenever required.

Temporal key Identifier (TKID): Device can have multiple PTK and GTK which are uniquely identified by TKID.

6.5.3 4-way Handshake

4-way handshake is used for authentication and generation of PTK using shared master key between pair of devices. Handshake is performed through security commands either transmitted using PCA or DRP.

Figure 6.2 shows 4-way handshake sequence diagram. Brief description of the different messages is as below:

Message 1: Initiator transmits PTK command as message 1, which includes MKID, proposed unique TKID, newly generated random number called initiator's nonce (I-Nonce). On reception of message 1, responder device checks uniqueness of TKID. If TKID is unique then it generates random number called responder's nonce (R-Nonce) and derives the PTK and Key Confirmation Key (KCK). Responder generates message 2 using this information.

Message 2: Responder transmits PTK command as message 2, which includes appropriate status code, newly generated random number called responder's nonce, PTK MIC of newly generated PTK and newly derived KCK. On reception of message 2 initiator device generates PTK, KCK, and recalculate the PTK MIC using KCK. If PTK MIC does not match or status

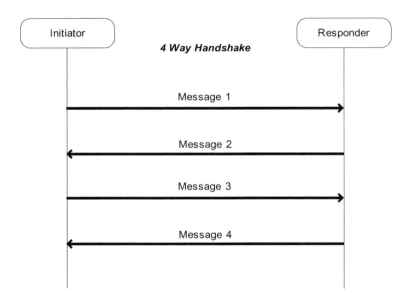

Fig. 6.2 4-way handshake.

code is not normal then 4-way handshake is aborted. Else, initiator generates message 3 using this information.

Message 3: Initiator transmits PTK command as message 3, which includes same initiator's nonce which was used in message 1 and PTK MIC computed using newly derived KCK. On reception of Message 3, responder device verifies PTK MIC using KCK. If PTK MIC does not match then 4-way handshake is failed. If PTK MIC matches then responder considers PTK as a temporal key with the initiator.

Message 4: Responder transmits PTK command as message 4, which includes same responder's nonce which was used in message 2 and PTK MIC computed using KCK. On reception of message 4, initiator device verifies PTK MIC using KCK. If PTK MIC does not match then 4-way handshake is failed. Else, PTK MIC matches then initiator considers PTK as a temporal key with the responder.

6.5.4 Distribution of GTK

Secure GTK commands are used to distribute GTK to the group of devices, using PTKs.

Initiator sends GTK command as message 1, with GTKID. On reception of message 1, device verifies the GTKID is unique TKID. If it is unique then sends message 2 with appropriate status code, else GTK distribution fails.

A recipient can request for GTK by sending command message 0. If initiator has requested GTK then it sends message 1 with GTKID and in response responder sends message 2. If Initiator does not have requested GTK then it sends message 1 without GTKIDID and request for GTK fails.

Previously distributed GTK can also be used for distribution of new GTK for group of devices.

6.5.5 Message Freshness and Replay Prevention

Each transmitter maintains a 48 bit secure frame counter per PTK and GTK. It is initialized to zero at start. Transmitter increments Secure Frame Counter (SFC) before transmission of the new frame or retry with the temporal key.

Each recipient has 48 bit replay counter per PTK and GTK. Similarly it is initialized to zero at start for PTK and same as GTK SFC for GTK, which can be found through GTK command during distribution. On reception of frame with SFC for the temporal key, recipient compares SFC with replay counter. If SFC is less than or equal to replay counter then that frame is discarded else frame is accepted. If frame is accepted then replay counter is set to SFC of the accepted frame.

6.5.6   Message Integrity

Frames are protected from modification by any intruders through authentication using Message Integrity Code (MIC). MIC is cryptographic checksum of the message to be protected. MIC also ensures the sender and recipient has same temporal keys, so it has gone through 4-way handshake and if required GTK distribution. AES-128 Cipher Block Chaining — Message Authentication Code (CBC-MAC) is used for MIC calculation. It also helps to prevent Man-In-Middle or sniffing attacks.

6.5.7   Payload Encryption

Data encryption is used to protect data from access by parties not having the encryption key. This key is a PTK for unicast traffic transmitted between two devices and a GTK for broadcast or multicast traffic transmitted by a sender to a group of recipient devices.

AES-128 counter mode is used for data encryption with PTK or GTK. Sender can even encrypt partial frame using encryption offset of secure MAC frame.

6.6   Ranging Support

Ultra Wideband device can provide support of one-dimensional ranging measurements between devices using Two-Way Time Transfer (TWTT) techniques. Support for range measurement is communicated through MAC Capabilities IE. It can be used for finding distance between two devices supporting range measurement feature. Supporting this feature is optional in the ECMA-368 specification. There are methods available where applications can use the ranging facility for device association (Ref. [25]).

Ranging uses PHY timer called range timer. Ranging is performed through three different commands (1) Range Measurement Request (RRQ) (2) Range

Measurement (RM) (3) Range Measurement Report (RMR). These three commands along with range timer are used for range measurement.

Device which is trying to find the distance (initiator) with another device (target), initiates the range measurement. Initiator turns on the range timer and transmits RRQ command to the target. On reception of RRQ, target supporting ranging feature enables its own range timer and transmits Imm-ACK to RRQ. This procedure is ranging initialization which is pictorially described in Figure 6.3.

Actual range measurement is done through range measurement commands. Initiator sends RM to target. At the time of transmitting the frame, initiator captures the time value from range timer after taking care of the calibration constant and adding transmit delay, let us call that value T-Initiator. On reception of RM command at target, the target also captures the time value from its range timer, taking care of the constants and reception delay, let us call that value R-Target. Target responds with Imm-ACK to RM command and captures that value as T-Target similarly to initiator. When initiator receives Imm-ACK, it captures the time as R-Initiator similar to target. Now, initiator and target both have two time values captured. Figure 6.4 shows range measurement procedure. This can be repeated multiple times for better results.

Range measurement report command is transmitted by target to initiator containing the T-target and R-target time values captured by target along with

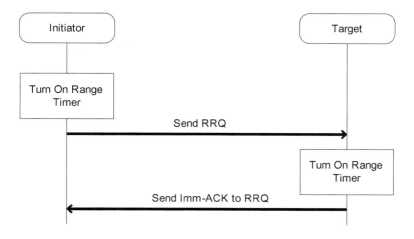

Fig. 6.3 Ranging initialization.

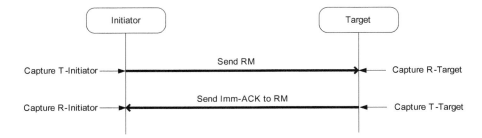

Fig. 6.4 Range measurement.

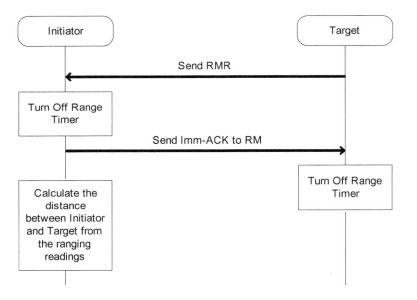

Fig. 6.5 Range report.

PHY clock accuracy (see Figure 6.5). Target provides multiple sets of these values if captured multiple times to initiator. On reception of RMR command, initiator can stop the range timer and responds with Imm-ACK for RMR. On reception of Imm-ACK target also stops the range timer.

The values from initiator, i.e. T-Initiator and R-Initiator and values from Target, i.e. R-Target and T-Target are now with initiator. Initiator uses these values to calculate the distance between initiator and target.

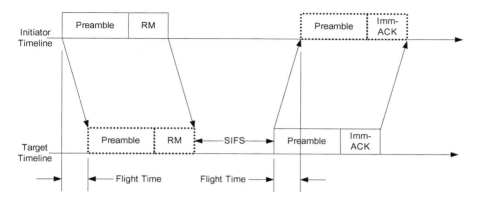

Fig. 6.6 Flight time.

Considering that both (Initiator and Target) the clock resolution are same, then two way flight time using these values is ((R-Target − T-Initiator) − (T-Target – R-Initiator)). Using only one way flight time the value is 1/2 ((R-Target − T-Initiator) − (T-Target − R-Initiator)). Figure 6.6 shows flight time concept in form of a picture.

Considering the signal propagation speed as speed of light, distance can be calculated as (Flight time × (speed of light/frequency of the range clock)).

The parameters like noise, multi-path effects, calibration constant accuracy, calibration constant precision, range timer clock resolution, range timer clock accuracy, timer clock difference between two devices and relative motion affects the accuracy of the distance measurement. So, it is good to take multiple range measurement for better results.

6.7 Multicast Address Binding

All multicast traffic uses multicast device address as destination address. This multicast must be mapped to multicast EUI-48 for upper layer usage. MAB IE provides binding information for multicast EUI-48 and multicast device address. Devices maintain the binding information internally and use those while transmission and reception of the frame to known bind multicast device address. Binding information is broadcasted through MAB IE in beacon on creation of binding and change in neighborhood so each device always has latest information available.

6.8   Transmit Power Control (TPC)

Device can transmit frames with different power but it has to be less or same as the power of its last beacon, so the signal does not go beyond the radio range of the device and harm any other communication. Device can suggest other devices about power level changes using Link Feedback IE. Similarly, it may get recommendation for power level changes through Link Feedback IE from its neighbors. Though, it is not mandatory to follow or send the recommendations. Device may use the signal to noise ratio, received signal strength, frame error ratio or other implementation specific parameters to decide the best transmit power. Device should use minimum possible transmit power required to maintain link with its neighbor.

6.9   Probe

Probe is used to get specific IE from neighbor device. Device can request one or more IEs through Probe IE (called implicit probe) or probe command (called explicit probe). Requested device can add the requested IEs in its beacon in next three superframes, if implicit probe. Requested device can send response through probe command in same superframe or in next three superframes, if explicit probe. Also, requested device can reject the probe (by Probe IE with no information element ids); or requested device may send response for subset of IEs requested.

Similarly, Application Specific Probe IE and Application Specific Probe Commands are used for requesting Application Specific IEs.

6.10   Multi-rate Support

Each device supports base rate of 53.3 Mbps. Beacons are transmitted at this rate. Information of supported rates is broadcasted through PHY Capabilities IE. Source device transmits the data at the rate which is supported by the target. Link Feedback IE from target device or any other implementation specific method can be used for rate adaptation.

6.11 Dynamic Channel Change

Dynamic Channel Change is a procedure, which allows a group of devices to change the channel in coordinated manner. Device can initiate channel change by adding Channel Change IE.

Channel Change IE includes:

- New Channel Number — New channel number for operation.
- Change Channel Count — the remaining number of superframes before the device moves to new channel.

The change channel count is decremented in successive superframes. Upon the change channel count reaching zero, the device moves to the new channel at the end of that superframe.

On reception of the Channel Change IE, a device that also intends to change channels in a coordinated manner includes a Channel Change IE with the same field values in its beacon and continues decrementing channel change countdown.

Selection of channel to move depends on system design and resource requirement of the device.

6.12 Aggregation

UWB MAC support unique feature of aggregation. A source can aggregate multiple MSDUs into single frame payload with only one MAC header and single identical delivery id field. On reception of an aggregated data frame, the recipient device delivers the separate MSDUs from the aggregated frame payload to its upper layer maintaining the ordering from the payload, like all MSDUs are received separately in correct order.

The aggregated MSDUs are aligned on 4-octet boundaries. Prior to each MSDU, 0 to 3 pad octets are inserted such that the MSDU starts on a 4-octet boundary.

6.13 Summary

In this chapter, we discuss advanced MAC features (besides interoperation and data transfer) which can make UWB system more secure and robust with long battery life. These features include acknowledgment policies, power save, security, and unique feature like range measurement. It also covers MAC support for transmit power control, release of unused reservations, probe, multi-rate support, dynamic channel change, and aggregation, which makes UWB system truly cooperative for better performance in WPAN environment.

7

UWB System Design

7.1 Introduction

Ultra Wideband system is quite different from other wireless connectivity systems, because of nature of UWB radio, expectation from the technology, varied use cases, and flexibility of implementations.

Ultra Wideband physical layer uses very low transmit power. The properties of the UWB physical channel and regulations of the usage necessitate some changes, especially in MAC and upper layer protocol implementations. UWB system design is expected to support various use cases in different forms. System implementation could be entirely hardware, single chip, multi chip, software plus hardware, firmware with operating system (OS), firmware without any OS, part of multi-interface chip etc.

We discuss some considerations for system design that can meet the expectation of desired environments.

7.2 System Requirements

Various applications and host environments affect system design for UWB connectivity system. This section discusses some of the system requirements of UWB system.

7.2.1 Throughput

Throughput is something which end user cares for the most. One can feel it with every file copy or with each video download. Higher throughput is desirable for any connectivity system. Many factors and system components are involved in throughput.

Throughput depends on hardware software interface data rate, processing speed of the processor of the host or processor of the System-on-Chip

(SoC), design of the hardware, software or firmware that determines how fast the system can process the received frames and frame transmission. Throughput also depends on host interface of the system, if it is connected to a host.

Overall throughput also depends on performance of upper layer and application. If any one component of the system performs badly, that can degrade the performance and throughput of whole system drastically. It can even lead to the buffer overflows, due to different data arrival and consumption rates, which eventually leads to system failure.

7.2.2   Software and Hardware Complexity

There are different ways of protocol implementation. These include hardware, software or combination of both. A good trade-off between hardware and software results into good performance with limited complexity, while allowing system maintenance and ease of upgrades. System is required to be upgraded for new variations of specifications. If complexity is handled in software then it is easy to upgrade.

7.2.3   Interface with Host

Depending on target applications and products, UWB connectivity system can have one or more different host interfaces. Host interface data rate must be fast enough to handle transmission rate and reception rate of the frames to achieve the optimal high speed system. It is possible that host interface is handled through software drivers. Performance of the interface and that of the system also depends on optimal design of interface drivers.

7.2.4   Multiple Protocol Support

Multiple upper layer protocol support is one of the most unique features of the UWB common radio system. Depending on where the system is used and what are the expected applications, system can provide the support for one or more PALs or implementation interface for multiple PALs. System must take into consideration the requirements from multiple PALs and manage the common radio such that many PALs can coexist. For example, if one PAL wants to switch off the radio for power saving and another is active at that time, then system must not switch off the radio. For multiple PAL system a decision making management entity is crucial, which can manage simultaneous active PALs without sacrificing different PALs requirements and overall system capabilities.

7.2.5 Future Compatibility

In fast paced technology world, time to market is a very important factor for the success of any product. Same is true for UWB products also. Products are expected to be future compatible and easily upgradable by software driver upgrade or firmware upgrade. System must be designed in such a way that hardware is quite configurable, so software or firmware upgrade is enough for future upgrade of the system. Hardware upgrade is comparatively costly and time consuming than software or firmware upgrade.

7.2.6 System Capacities

It is possible that some of the system capacities and capabilities of UWB system are hard bounded during implementation, like maximum number of neighbor devices, number of active reservation at a time, length of the queue of the frames to be transmitted, number of active PALs at a time etc. These capacities are defined based on environment, where device is supposed to be used and intended applications over radio platform. The design must handle the overflow condition of these kinds of capacities without creating any on-air violation of the protocol behavior and resultant failures. One of the valid solutions for any unavoidable scenario handling is graceful shutdown of the system.

7.2.7 Being Good Neighbor

Each device is expected to be a good neighbor in wireless connectivity ecosystem, so that multiple devices can coexist with different kind of applications and traffic requirements. Good neighbor is expected to provide room for new devices and applications. MAC defines many rules and policies to achieve this goal for UWB WPAN. Even then devices can try to be greedy in using resources and hog the channel with reservation in excess of what is required. It is always strongly suggested not to reserve more than required for the application. Reservations can be modified any time for increase or decrease in resources based on traffic requirements. Being good neighbor could be showcase feature for the product.

7.3   Hardware Software: Partitioning and Co-design

Considering different requirements of the UWB system, the most common system design can be based on hardware software partitioning. Time critical functions of the system can be implemented in hardware for real time performance without fail. Functions, which are not time critical are generally implemented in software or firmware.

Whatever is implemented in hardware is not easy to change because it requires re-spin of the chip, which is very costly and time consuming. While PHY features are generally implemented in hardware, MAC features can be partly implemented in the hardware and partly in software to ensure good trade-off between complexity and upgradability. MAC specification is designed in such way that time critical functions of the system, like frame formats, time intervals etc. does not change during change in specification. It makes system upgrade easy without change in hardware. Aligning to this philosophy, system also must be designed such that time critical functions which are not expected to change go in hardware to provide consistent real time performance.

Each transmitted frame includes MAC frame header attached to it. Frame header generation being a repeated task could either be in hardware or software. This decision can be taken based on many factors. Number of MAC frame header generation depends on arrival rate of data for transmission and fragmentation limit set for the transmission. Other factors to consider are the capability of processor to generate the frames of transmission and availability of the memory. On other hand, reception depends on external environment and capability of peer device, data rate supported by the peer device, size of the fragments etc. Due to external factors, worst case needs to be considered for reception. Device must handle highest possible data rate supported, and arrival of maximum possible number of frames in a time unit. Reassembly of frames must be fast enough so that it can handle new frames received without fail in limited buffer depending on system buffer availability. Further, depending on speed of the processor and hardware interface, received frame header processing and reassembly mechanism are implemented in hardware or software.

Also, MAC upper layer must consume the MSDU soon; otherwise buffer size required is very large.

On reception of frame with acknowledgment policy set to Imm-ACK, an Imm-ACK frame is sent after SIFS duration. SIFS is very small time duration. To meet this time requirement, Imm-ACK mechanism is generally implemented in hardware. Frame with B-ACK request acknowledgment policy also requires B-ACK transmission in SIFS, but protocol provides way to send response later, so can be implemented in conjunction with software and hardware. RTS and CTS frame also have similar timing requirements so generally implemented in hardware. Similarly, slot time of PCA is a tight timing requirement so generally back-off countdown is implemented in hardware.

Synchronization is a very important feature for the UWB system. It is based on reception time of beacon. For proper synchronization, beacons must be transmitted at beacon slot boundary. Synchronization can take care of internal system timing errors. It can even use historical measurement of synchronization for better performance. System design must take care of these requirements of synchronization. Based on actual on-air tests, synchronization algorithms can be optimized for best performance using software or firmware.

Queuing of traffic based on different access category or different stream can be either done in software or hardware. It mostly depends on how much configurability is expected from the system. If upper layer is going to use only few subset of traffic access category or only single stream then it is better to implement queuing in hardware. But if all the access categories and multiple streams are supported then lots of dedicated memory is required for multiple queues. In that case, it is recommended to implement this in software.

Power usage is a very important feature for any handheld wireless system. MAC provides protocol and defines on-air behavior for power save features. PHY also has its own power save mechanism. Some level of power save can be achieved by transmit power control as well. Actual system power save can be implemented in hardware, which can be configured through software. Power save is implementation dependent, and therefore could be a very novel design by itself for the system complying on-air behavior. This can be a differentiated feature amongst multiple product offerings.

There are many other features like rate adaptation, transmit power control, and link feedback usage etc. implemented in many different ways which can improve the performance of the whole system.

7.4 Common System Architectures

Some trade-offs for UWB system design based on requirements and impact on decision of hardware and software partition are discussed in the previous section. Overall system including PAL(s) can be designed in different ways. Factors affecting system architectures include upper layer application or protocol, the container of the UWB system, i.e., where the UWB system is going to be used in mobile handheld, PC, camera, printer etc., and host interfaces.

Followings subsections briefly describe few common system architectures mainly depending on host interfaces and system component arrangements.

7.4.1 Hardware Host Interface

Figure 7.1 shows the PHY and partial MAC in hardware and it is connected to host through some interface; this host interface usually is PCI, PCMCI etc. Time critical elements of MAC are implemented in hardware. Complex MAC protocol and state machines are usually implemented in software with software interface for upper layer as part of drivers on some operating system.

Host processor needs to process MAC protocol and all transmitted and received frames in this architecture, so host processor is really loaded with

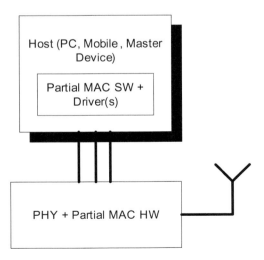

Fig. 7.1 Hardware and host interface.

MAC processing. Products with this architecture type could be in the form of PCI or PCMCI cards.

7.4.2 System-on-Chip or On Board Processor

Figure 7.2 shows the PHY and partial MAC in hardware. Partial MAC is implemented in the form of software or firmware running on chip processor or on some dedicated processor. The hardware MAC and software MAC should have very fast communication link. This system is connected to host through some interface and can be operated by thin driver on host.

Here, most of the complexity of the protocol is handled in hardware and software on dedicated processor, so it does not load up host processor much. Products with this architecture type could be in form of PCI or PCMCI cards, with SoC having PHY, hardware MAC and on chip processor running software part of the MAC.

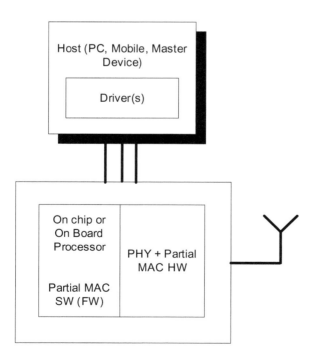

Fig. 7.2 Hardware host interface with on chip or on board processor.

7.4.3 Complete Hardware System

Figure 7.3 shows full hardware system and it has direct interface with host. Here all the MAC protocol implementation is done in hardware and host has simple thin driver to handle that. This architecture is only used in case if system is used for dedicated application and implementing minimal part of MAC for its operation, so hardware remains relatively less complex.

It is possible to have hardware PHY and partial MAC in single chip or PHY can be a separate chip (ECMA-369 defines MAC–PHY interface also) and partial MAC or full MAC could be another chip. PHY and MAC hardware can be connected through high speed bus interface.

These are some common architecture types, but other architectures are possible based on actual use case of the UWB system.

If the system is dedicated for only one kind of application then it is possible to have full hardware system without any firmware or software and without any host interface. Or the architecture may include system without any host interface and with an on chip processor doing most of MAC functionalities with partial MAC in software.

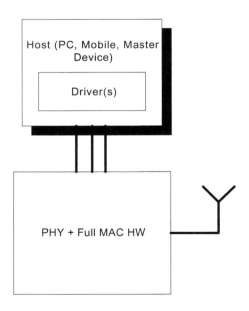

Fig. 7.3 Full hardware system.

System which caters multiple upper layer protocols has more complex and extendable architectures with many components in firmware or software. These systems should have the implementation of specific radio management entity to support different upper layers requirement and radio resource; so it can manage internal contention for the radio resources. Chip vendors provide multiple different host interfaces like PCI, PCMCI, USB etc. for the system which can be used for various applications and environment.

7.4.4   USB Wire Adaptors

This is generally used by Wireless USB system, where host interface is wired USB. Host has drivers for interface incorporating many MAC protocol related complexity and either hardware or software on dedicated processor acts according to instructions from class drivers. Details of the wire adaptors are described in Chapter 9. Even other PALs can be implemented with USB being host interface.

7.5   Interoperability Considerations

Interoperability is one of the key factors for the success of wireless products. Wireless device needs to coexist with similar devices from multiple different vendors.

Standards provide room for many different options for feature implementation, and vendor specific implementation of protocols. It also provides room for optimization of algorithms for better performance with many different PAL implementations. Standards are defined to promote interoperability between devices. It is possible that two devices may not communicate with each other but they can operate in proximity without creating any problem for each other.

Beaconing protocol is the foundation for interoperability. It provides basic infrastructure for distributed UWB MAC so that devices implementing ECMA-368 specification can coexist. Even application specific IEs, application specific commands, and application specific control frames that are used for various vendor specific requirement, are defined to allow interoperability.

In case some of the optional features of the PHY and MAC are not supported then it can be informed through MAC and PHY Capabilities IE. The device receiving information about MAC or PHY capabilities from a neighbor device, should not exercise unsupported feature of the neighbor device.

This leads to good interoperation and does not waste the bandwidth of the channel. In case device exercises peer device's unsupported features then peer device may not react to the request or data. The requesting device should be capable of handling that failure, and must not retry indefinitely. Otherwise, it can lead to poor interoperability and system deadlock.

It is also possible that implementation has some inherent limitations like maximum number of devices in neighborhood, maximum number of reservations etc. If more resources are required, then system should either shut down or reject more requests without breaking on-air behavior of the MAC protocol.

Beaconing is the heart of UWB MAC system. It is required that beacon is received by all neighbor devices. Reservation negotiation and maintenance is done through beacons and traffic indication map is communicated through beacons. Beacons are transmitted with power level equal to or higher than transmit power of data. For better interoperation it is recommended that the beacons are sent at highest allowed transmit power, so if required data also can be transmitted with highest power due to change in channel conditions and mobility.

Some PAL may not require support for PCA access method or some type of reservations. In this case, it may not participate in PCA or unsupported reservations and data transfers in those time periods. But, it is possible that PCA or unsupported reservation exists in neighborhood. System must handle variations in neighborhood without failure of interoperability.

Private reservation is a special kind of reservation, which has its own application defined channel access method that can be used in vendor specific protocol design. Multiple PALs are required to select correct security mode depending on the requirement of the PAL to ensure the security of multiple PAL sharing common radio and one single beacon for the common platform. Beacon indicates the security mode supported by the device.

In summary, standards define basic features, and good implementation design leads to a good interoperable UWB system.

7.6   Product Certification

With interoperation, devices are also expected to implement the protocol in correct fashion and act as a good neighbor. To ensure these requirements, WiMedia Alliance provides the certification for the products which are based on UWB common radio platform, implementing ECMA-368. The WiMedia Certification Program enables the complete coexistence and interoperability of radio platform, different end applications and protocols based on approved UWB standards. These certifications include PHY certification, and platform certification. This certification ensures that the end product can coexist and interoperate regardless of the vendor of the product.

ECMA-369 defines interface for the PHY and MAC, it allows vendor to implement PHY only or MAC only. MAC and PHY of any different vendors can be used for system development through standardized MAC–PHY interface. Vendor even can implement optimized system with single chip MAC and PHY, without any standardized PHY interface. Certifications of PHY and platform only mandate the on-air behavior of the system.

PHY vendors can certify only PHY through PHY Certification Program. While UWB common radio platform with MAC and PHY can be certified under platform certification, which requires PHY to be certified that is used in the platform.

The certification process has been put in place by WiMedia Alliance, to ensure that all products in different forms (PHY chip, radio platform) with different PALs from different vendors can coexist, can interoperate and be good neighbors to each other.

7.7 Summary

Ultra Wideband system requirements were discussed in this chapter. Their impacts on design and performance of the system are highlighted. Complexity of the system, application's usage, and usage environment of the UWB system needs to be considered for high performance UWB system design.

Common system architecture and container of different system modules like PHY, MAC, interface drivers, firmware were described in this chapter. There are different possibilities for host interface with connectivity system of UWB and different possible ways to design the whole system.

For best outcome of the system, interoperation and coexistence is very important requirement and standard certification provides ensured interoperability between different vendor devices.

8

Adaptation to Multiple Applications

8.1 Introduction

Ultra Wideband common radio, name itself describes that the radio can be used for multiple upper layers and applications. More than one upper layer can work simultaneously above Medium Access Control (MAC) layer, without disturbing functionality of each other. At the MAC layer, frames are transmitted according to MAC frame format and rules. MAC provides different ways to multiplex and de-multiplex data traffic to appropriate upper layer or application. Common radio architecture is such that, it is possible to write a direct application or proprietary protocol layer above MAC for data transfer. It also has the facility for defining proprietary protocol for control mechanism using Application Specific IEs, Application Specific Control frames, and Application Specific Command frames. Wireless USB, WiMedia Link Protocol (WLP) etc. are some of the known upper layers which uses facility of UWB common radio platform for their operation.

This chapter covers the unique tools provided by UWB common radio platform for adaptation of multiple applications and upper layers. These tools are the ones that use the MAC protocol to make highly extendable UWB common radio platform.

8.2 Traffic Multiplexing Methods

Medium Access Control does not understand the difference between different upper layers or upper applications. MAC receives frames from upper layer in form of MAC Service Data Unit (MSDU). MSDU is fragmented, if required and passed to PHY for transmission in form of MAC Protocol Data Unit (MPDU).

Finally for UWB system, traffic from upper layer must be multiplexed at MAC layer for transmission and de-multiplexed at reception to reach correct upper layer and application.

Medium Access Control defines different mechanisms for handling link to upper layer. Traffic from different upper layers like WUSB and WLP can be multiplexed in different ways as follows.

8.2.1 MUX Sublayer

Traffic can be routed through MUX sublayer between MAC and application protocol or protocol adaptation layer. Upper layer frame is attached with a MUX header which provides the identification of frame (which is indeed an identification of upper layer also). Resulting upper layer frame and MUX header forms MSDU for MAC.

Frame format for MUX sublayer is shown in Figure 8.1. On transmission side MUX header is attached to each MSDUs entering to MAC according to its application or upper layer. If required, formed MSDU is fragmented and multiple MPDUs are transmitted by MAC.

On reception side MUX header is removed from the reassembled MSDU and actual upper layer frame is routed to appropriate application layer according to MUX header. Figure 8.1 shows the MSDU with MUX header and payload.

MUX payload field is actual frame from the application or upper layer.

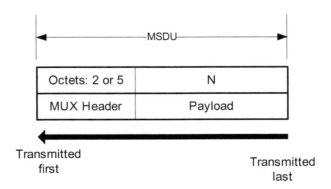

Fig. 8.1 MSDU with MUX header.

MUX header is variable length. There are 2 different variations of MUX header. Depending on first 2 octets of the MUX header, type of MUX header is identified.

If value of first 2 octets is between 0x0000 to 0x00ff then it is 5 octet long MUX header as shown in Figure 8.2. This is called Organizationally Unique Identifier (OUI) based MUX header. If value of first 2 octets is between 0x0600 to 0xffff then it is 2 octet long MUX header as shown in Figure 8.3. This is EtherType based MUX header.

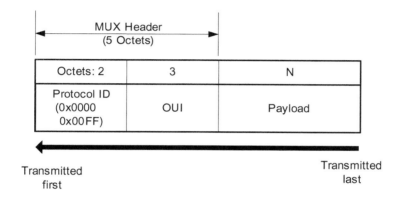

Fig. 8.2 MUX header (5 Octets).

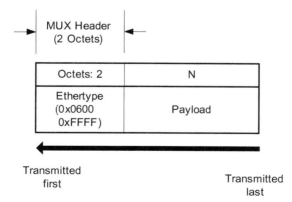

Fig. 8.3 MUX header (2 Octets).

8.2.1.1   OUI based MUX Header

Organizationally Unique Identifier is 3 octets unique number that can be purchased from IEEE. This number uniquely identifies vendor, manufacturer or organization worldwide. MUX header with first 2 octets from 0 to 255, has unique 3 octets OUI followed by 2 octets.

Here, first 2 octets define Protocol ID. The Protocol ID field can be set to values from 0 through 255 and is set to a value that identifies a protocol defined by the owner of the OUI specified in the OUI field.

8.2.1.2   EtherType based MUX Header

EtherType is 2 octet field. The EtherType field provides a context of upper layer for interpretation of the data field of the frame. Well known protocols already have an assigned EtherType value, which is managed by IEEE. For MAC upper layer protocol EtherType value starts from 0x0600.

If the first 2 octets of MUX header have value greater than or equal to 0x0600 then it is EtherType based MUX header. EtherType field can be set to values from 1536 to 65535. Each value of EtherType identifies an upper layer protocol for MAC.

8.2.2   Reservation based Multiplexing

Applications or upper layers can use a reservation for multiplexing and de-multiplexing of upper layer traffic. Each reservation has stream index, source address and destination address to multiplex and de-multiplex outgoing and incoming traffic. Regular MAC header is used for traffic sent using DRP with stream index, source address, and destination address associated application protocol. Figure 8.4 explains the reservation based traffic multiplexing and de-multiplexing graphically.

8.3   Application Specific Frames

Medium Access Control defines different application specific frames for usage in new upper layer or proprietary protocol. For all application specific usage, 2 octets field is defined called specifier ID. Vendor who wants to implement some vendor specific protocol has to get a unique specifier ID from ECMA.

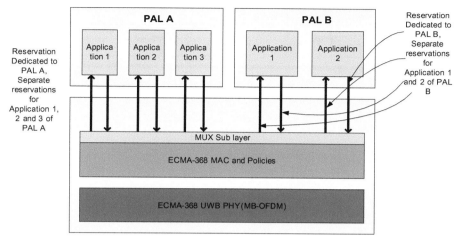

Fig. 8.4 Reservation based multiplexing.

This ID gets listed as unique specifier ID for the vendor by ECMA as UWB specifier ID for the vendor. The owner of the specifier ID can define its own payload for the frame with that specifier ID. The frame with that specific specifier ID and its payload content is understood by only those devices which implement same upper layer protocol or proprietary protocol. Other devices do not understand the specifier ID and ignore these frames. This provides the extensibility for different kind of protocol implementation.

Following are different application specific frames defined for application specific usage.

8.3.1 Application Specific IE

Figure 8.5 shows application specific IE. It is transmitted as part of beacon frame by request of upper layer or for some proprietary protocol identified by 2 octets specifier ID. This IE is used for information broadcast by owner of specifier ID. Element ID for Application Specific IE is always 0xFF. Format of application specific data is defined by the owner of the specifier ID. It may include information related to multiple protocols as part of application specific data under one allocated specifier ID.

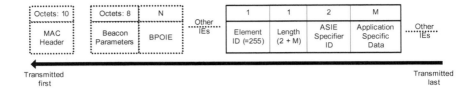

Fig. 8.5 Application specific IE.

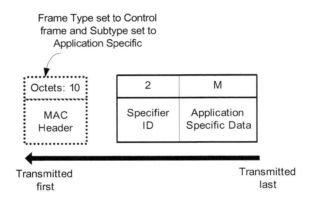

Fig. 8.6 Application specific control frame.

8.3.2 Application Specific Control Frame

Figure 8.6 shows Application Specific Control Frame. It is transmitted with MAC header same as any other control frame, with frame subtype set as Application Specific Control frame. It starts with 2 octets specifier ID. Here also, format of Application Specific data is defined by the owner of the specifier ID. Devices who implement the protocol with this specifier ID respond to these Application Specific Control frame accordingly. Having one specifier ID for a vendor, the vendor can define the application specific data in such a way that it can be used by many protocols.

8.3.3 Application Specific Command Frame

Application Specific Command frame is command frame similar to Application Specific Control frame. Figure 8.7 illustrates the frame format. It is also understood by and transmitted by the devices which implement the protocol identified by the specifier ID.

Frame Type set to Command
frame and Subtype set to
Application Specific

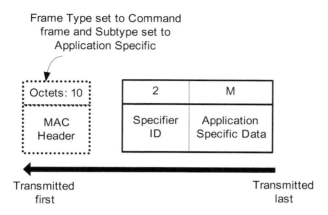

Transmitted
first

Transmitted
last

Fig. 8.7 Application specific command frame.

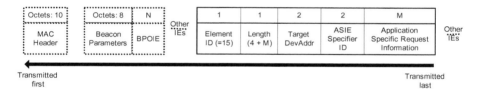

Transmitted
first

Transmitted
last

Fig. 8.8 Application specific probe IE.

8.3.4 Application Specific Probe IE

A special kind of probe IE is defined to probe the Application Specific IE for the device supporting a defined specifier ID. Figure 8.8 shows Application Specific Probe IE. Again, content of IE starts with destination address and specifier ID, which is only understood by the device implementing protocol specified by the specifier ID, other devices ignore it and do not respond.

8.4 PCA Access Category Mapping

Prioritized Contention Access (PCA) provides differentiated and distributed contention access method. It supports 4 access categories for transmission of data. Depending on access category, device has different transmission back-off and size of contention window. Ultimately, the traffic with higher priority and QoS requirement has priority for transmission over wireless medium.

Table 8.1 User priority to access category mapping.

Priority	802.1D User priority	802.1D Traffic type	PCA Access category	Traffic
Lowest	1	Background	AC_BK	Background
	2	Unused	AC_BK	Background
	0	Best Effort	AC_BE	Best Effort
	3	Excellent Effort	AC_BE	Best Effort
	4	Controlled Load	AC_VI	Video
	5	Video	AC_VI	Video
	6	Voice	AC_VO	Voice
Highest	7	Network Control	AC_VO	Voice

Upper layer protocol also includes the information about packets QoS requirements. It is called prioritized QoS. IEEE 802.1D defines the mapping of use priority and traffic type. These user priority and traffic type can be mapped to four access category of PCA in ECMA-368, so PCA can provide priority to that traffic. Table 8.1 shows mapping of user priority to PCA access category.

8.5 Upper Protocol Layers

While UWB standards were under development, many other industry alliances acknowledged advantages of UWB common radio. UWB is expected to provide high speed wireless communication medium for short range and have unique capabilities to support multiple upper layer protocols. Taking these into account, different alliances simultaneously worked toward defining upper layers adaptation layer of different existing legacy protocols over UWB radio. Some industry alliances adopted ECMA-368 based UWB as radio layer for its protocol layer. Following subsections brief some such popular protocol adaptation layers using UWB common radio for adaptation of different existing protocols of wired connectivity origin.

8.5.1 Wireless USB (WUSB)

Wireless USB is one of the first adopters of UWB as its radio technology. WUSB provides adaptation layer for legacy USB to work wirelessly over UWB high speed radio. It uses private reservation for all its operation. All incoming and outgoing traffic during WUSB private reservation is sent to WUSB PAL. No WUSB packets are transmitted outside this private reservation. A Specifier

ID is defined for WUSB to use it for PAL specific control frames. Chapter 9 describes Wireless USB in detail.

8.5.2 Internet Protocol

Internet protocol is widely used protocol in PC, consumer electronics and other handhelds. Usage of internet protocol over wireless is also very well known. Internet protocol over UWB is expected to target home networking and the home entertainment environment. UWB radio can provide very high speed wireless Internet protocol. Internet protocol adaptation layer over UWB common radio is defined as WiMedia Link Protocol (WLP). WLP has capability of bridging frames from UWB and WLP PAL to Ethernet and other networking protocol. It uses reservation based as well as contention based medium access methods to support different QoS requirements of TCP/IP traffic. It supports parameterized QoS using reservation based mechanism and uses contention based mechanism for prioritized QoS requirements.

8.5.3 Wireless 1394

Wired IEEE 1394 was built for supporting high throughput multimedia services. 1394 Trade Association also considered UWB for its high speed wireless replacement of wired connectivity and proposed Wireless 1394 PAL. This PAL permits IEEE 1394 devices and protocols to be used in a wireless environment at a very high data rate of UWB common radio protocol, while allowing compatibility with existing wired 1394 devices. This PAL tries to support all IEEE 1394 services using UWB common radio facility and traffic capabilities of common radio. Wireless 1394 devices can coexist with other UWB devices with forming virtual 1394 channel and network.

8.5.4 Bluetooth over UWB (BToUWB)

Bluetooth has become omnipresent short range wireless connectivity solution. Today, for example, it is difficult to imagine a mobile phone without Bluetooth connectivity. Benefits of UWB in terms of very high data rates are encouraging the Bluetooth Special Interest Group (SIG) to consider its adoption. This is further facilitated by the UWB common radio platform.

Bluetooth SIG initiated work to adopt ECMA-368 based UWB for its high speed alternate PHY, while maintaining backward compatibility with existing Bluetooth framework. This is further discussed in Chapter 10.

8.6 Summary

This chapter provides the information for the users of UWB common radio platform. UWB common radio platform as name says is extendable can be used by multiple protocol adaptation layers and various applications.

Medium Access Control provides different tools and hooks which can be used by PAL or upper layer application to perform various tasks which is out of MAC's responsibility.

Medium Access Control provides multiplexing method through MUX header and reservation. Protocol Id or EtherType is used for multiplexing in MUX header. Any of these combination of multiplexing methods can be used by upper layer.

Also other various command and control frames, and information elements are defined for application specific usage. ECMA maintains specifier IDs that are used to distinguish those frames.

There are different PALs existing and efforts are ongoing to define new PALs over UWB common radio platform. Some of those are like Wireless USB, Internet protocol adaptation, Wireless 1394, BToUWB etc.

9

Wireless USB

9.1 Introduction

Universal Serial Bus (USB) has been very popular connectivity medium for PC and devices connecting to PC. Besides PC, today USB has become integral part of many consumer electronics devices like digital camera, music players etc. Wireless USB has emerged to unwire the well known USB. Wireless USB (WUSB) is suitable for many different devices like flash drives, external hard drives, printers, scanners, digital cameras, music players etc. High speed wireless data rates enable audio and video streaming. In addition, WUSB brings mobility to the USB devices.

In this chapter, we discuss the transformation of USB into Wireless USB by employing UWB common radio platform. For description of wired USB, readers may refer [97].

9.2 Evolution of Wireless USB

In current communication era, different digital-medias are converging with each other. As discussed in earlier chapters, consumer electronics, mobile communication and PC medium are interplaying with each other like never before. Common interconnection standard is necessary for proliferation of these mediums. USB has been preferred interconnectivity solution for these mediums taking into account requirements of throughput and latency. Evolution of very high speed short range wireless in form of UWB is contributing to WUSB, to emerge as very suitable connectivity for all of these requirements (similar to wired USB) without wires.

Consumer electronics market always has expectation of wireless interface with high throughput because of high quality content transfer required amongst the devices. Figure 9.1 shows that in a CE use case, a streaming device can

Fig. 9.1 Consumer electronics use case.

stream high quality media content to multiple devices (e.g. display device, recorder etc.).

Even mobile phones have support of USB connectivity, which can be used for connecting mobile phone to PC and other USB supporting devices. WUSB can be used to unwire mobile phone's USB connectivity. Figure 9.2 shows a representative set of mobile devices that can be connected through Wireless USB.

Today, it is unimaginable to visualize PC, laptop or notebook computers without USB ports. As WUSB is expected to be wireless replacement of wired USB, it can be used to wirelessly connect all devices which are today connected through USB. Figure 9.3 shows some examples of WUSB connectivity in PC domain.

9.2.1 Wireless USB and Ultra Wideband

The Wireless USB Promoters Group is chartered with defining the WUSB specification. WUSB Promoters Group has decided to adopt UWB common radio based on ECMA-368 for high speed wireless connectivity for WUSB. Wireless USB follows similar architecture as wired USB, but UWB

Fig. 9.2 Mobile use case.

Fig. 9.3 PC use case.

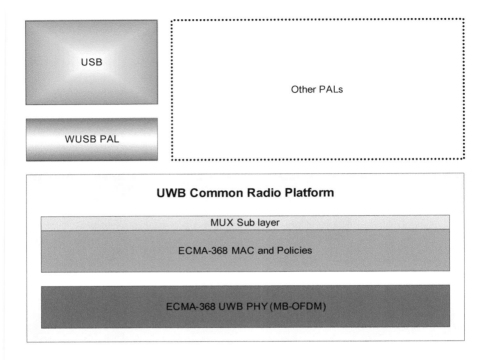

Fig. 9.4 UWB common radio platform with WUSB.

radio is placed in place of wired connectivity medium. This enables almost seamless migration of USB applications over WUSB. WUSB provides adaptation to UWB radio through protocol adaptation layer. Figure 9.4 shows the WUSB protocol with UWB common radio platform. MB-OFDM based UWB's capability of providing data rate up to 480 Mbps enables high data rate for WUSB.

9.3 WUSB Protocol

Wireless USB replaces USB such that application data transfer is identical to USB data transfer.

Similar to USB, WUSB is also classified in WUSB host and WUSB device. WUSB devices can be quickly connected to WUSB host, configured, used, and disconnected.

Wire in USB serves as means of association, but WUSB has different association models defined for associating a device to the specific host in absence of wire.

Wire also provides inherent protection for data because of closed wired medium, but wireless does not. WUSB uses security mechanism provided by MAC for security purposes.

UWB common radio at MAC layer is originally decentralized network topology but WUSB is centralized network. The MAC feature (beacon protocol) enables coexistence of WUSB host and devices with any other UWB devices (based on ECMA-368) around. Details of MAC features are described in Chapters 5 and 6.

9.3.1 Topology

Wired USB is based on centralized bus architecture, with host acting as master and device as slave entity. Host and device(s) are electrically connected to each other. Similar to wired USB, wireless connectivity of WUSB allows multiple device connections with WUSB host. Figure 9.5 illustrates WUSB topology, where multiple WUSB devices are connected to central host.

The WUSB host can connect up to 127 WUSB devices. WUSB host and connected devices are collectively called WUSB cluster. More than one WUSB cluster can coexist in overlapping spatial environment in same or different channel, because of UWB common radio platform and underlying decentralized MAC. Similar to USB, a WUSB host communicates with multiple WUSB devices, and a WUSB device connects and communicates with a single WUSB host. Topology also supports dual role model where a device can also support host capabilities.

Proprietary implementations based on UWB common radio platform are possible where WUSB hosts can communicate with each other directly and extend the capabilities of traditional USB (Refer [77]).

9.3.2 USB to WUSB

Evolving from USB to WUSB, changes in architecture are minimal. As a result, USB applications can seamlessly work on WUSB.

Figure 9.6 shows the changes in USB architecture for making it wireless.

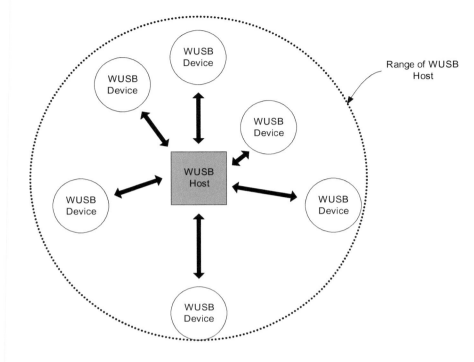

Fig. 9.5 WUSB topology.

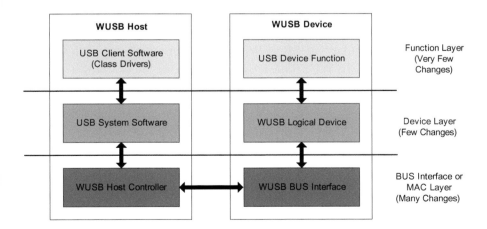

Fig. 9.6 WUSB architecture.

Wireless USB must provide the adequate bandwidth performance to meet typical user experience same as USB. UWB radio provides physical layer data rate up to 480 Mbps which is comparable to throughput of wired USB 2.0.

Similar to USB, WUSB devices are also expected to be near to the host. UWB is a WPAN technology and radio provides connectivity range up to 10 meters which is aligned to WUSB's requirement.

Universal Serial Bus is wired medium so device is powered up from wired connection. WUSB device does not have any wire connection with the host, so WUSB device is required to be battery operated. It makes power an important consideration for WUSB. Couple of day's battery life is a typical requirement for highly mobile devices. Battery life for the devices like remote control, which is not used constantly, should be in months. Therefore, the WUSB is required to be very low power. UWB radio is low power radio so long battery life can be achieved through UWB technology.

In addition, following are major changes introduced to transform wired USB into Wireless USB.

- WUSB does not have inherent security of wire, so new states and mechanisms for security are added in USB framework.
- WUSB has new descriptors added, some of the descriptors and endpoints are changed, which helps in making wired USB to wireless.
- WUSB clubs IN and OUT transfers of USB to reduce Rx-Tx switching of radio, by the help of a new command, Micro-scheduled Management Command (MMC).
- Device notification command is added for notification from WUSB device to WUSB host.
- The WUSB host has an identification name or number. This number is used at connection time by the device.

Similar to USB, WUSB also defines different types of devices as described below:

9.3.2.1 WUSB Host

Wireless USB host is a central entity for the WUSB ecosystem. There is only one WUSB host in WUSB cluster. WUSB host connects to host processor (generally PC or laptop) through interface called WUSB Host Controller Interface (WHCI) or through USB.

Wireless USB host manages connected devices. Channel access, controlling of devices, data transfer, power management, etc. are controlled by the host. Host supports all MAC features because it has to coexist with multiple WUSB cluster in same spatial area and with other UWB devices (non-WUSB also) in its radio range.

9.3.2.2 WUSB Device

Wireless USB Device is similar to USB device and connects to WUSB host. Device follows the host's instructions for channel access, connection, data transfer, power save etc.

Wireless USB device address is given by the host on enumeration. Device supports one or more pipes for communication with the host. Each device has a control pipe through which it performs the control command exchange with the host.

To coexist with other UWB devices, WUSB device has to be a good neighbor supporting MAC features. Depending on MAC support, WUSB devices are classified as follows:

Self Beaconing Device: It supports complete beacon protocol. Self beaconing device receives the beacons from all neighbors, parses those and transmits its own beacon according to beacon protocol. Figure 9.7 shows how Self beaconing device takes care of neighboring UWB devices with beacons.

Directed Beaconing Device: This device does not implement MAC features, but device behaves according to MAC rules on the air. MAC operations for directed beaconing device are performed by the host. Directed beaconing device captures the UWB frames (including beacons) without processing those. These frames are sent to the host. Host parses the frames received by the device and generates the beacon to be transmitted by the directed beaconing device. This beacon is sent to the device, and device transmits it at required time as informed by the host. On-air behavior of the device is same as Self beaconing device, but all MAC processing is done by the host for the device. In this case, device becomes very simple in implementation, but on-air traffic increases because of information exchange between host and device. Support

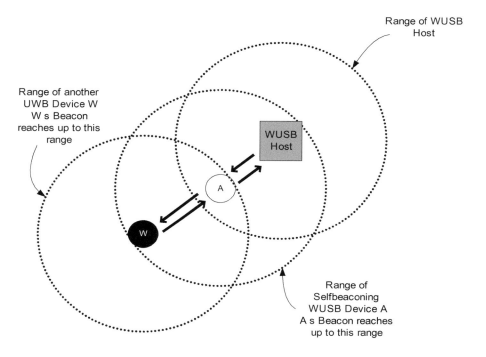

Fig. 9.7 Self beaconing WUSB device.

for directed beaconing device by the host makes host comparatively loaded and complex.

Nonbeaconing Device: Nonbeaconing device does not implement MAC protocol and also does not transmit any beacons. Transmission power and receive sensitivity of Nonbeaconing device is reduced such that range of this device is not more than half of the host's range. Due to the reduced radio range of the device, it always exists very near to host. Figure 9.8 shows this scenario. None of the signal from device can reach to any other UWB devices out of host's range. Nonbeaconing device keeps implementation of device as well as host simple.

9.3.2.3   Dual Role Device

Dual Role Device (DRD) is a WUSB device, which can act as WUSB host also. WUSB DRD can connect many other devices and can connect to a host being

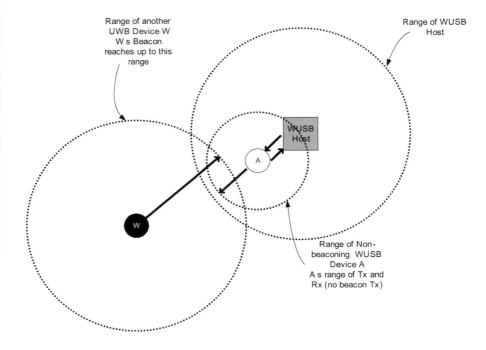

Fig. 9.8 Nonbeaconing WUSB device.

device. It enables new kind of scenarios like combination and point-to-point communication. Figure 9.9 shows topology that can be created by DRD.

9.3.2.4   Wire Adapters

Wire adapter delivers Wireless USB capabilities to existing wired USB host or device. The device that is used to connect wired USB devices to WUSB host is called Device Wire Adapter (DWA). WUSB host which connects to host processor via USB interface is called Host Wire Adapter (HWA). WUSB defines special protocol for support of wire adapters. Figure 9.10 shows HWA and DWA.

9.3.3   WUSB Frame Format

Wireless USB frames are based on MAC frame. General structure of WUSB packet contains PHY preamble, PHY header, and MAC header. MAC header

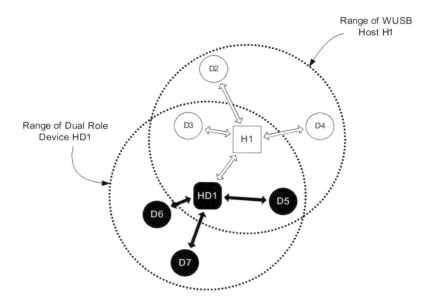

Fig. 9.9 Dual role device.

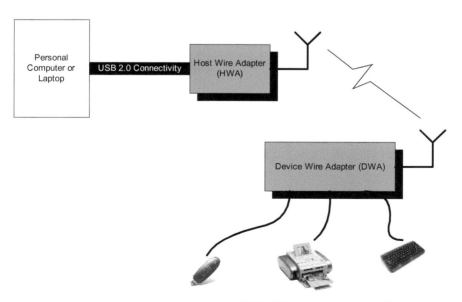

Fig. 9.10 Wire adapters.

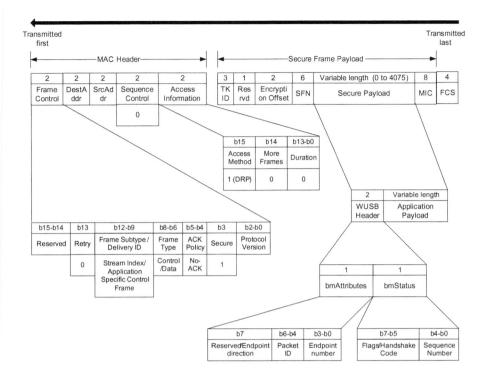

Fig. 9.11 WUSB frame format.

and payload fields are used to create different types of WUSB frames. Figure 9.11 explains detailed WUSB frame format with MAC frame field's settings for WUSB.

After secure relationship is established between host and device, WUSB frame is always a secured frame with secure bit set in MAC header frame control.

Wireless USB uses MAC control and data frame so frame type field is set from either of these. It uses only Application Specific Control frame from MAC control frames.

Wireless USB uses private reservation of MAC for its operation. WUSB host is owner of the private reservation. In case of data frame, delivery ID is set to the stream index chosen by WUSB for WUSB private reservation (also called WUSB channel).

Wireless USB has its own acknowledgment, retry, and sequence control. As a result, retry bit is always set to 0, acknowledgment policy is set to NO-ACK and sequence control is set to 0.

Wireless USB uses private reservation only, so access method is set to DRP in MAC header.

9.3.4 WUSB Addressing

Wireless USB host and devices use address fields of MAC header for WUSB addressing as well. Different address ranges are defined by MAC for different purposes, as discussed in Chapter 5. Similarly, WUSB defines different address range for its internal operations.

WUSB Host Addressing: WUSB host has unique EUI-48 address. It uses generated unicast MAC device address (range between 0x0100 to 0xFEFF) in its beacon as well as same address as source address in transmitted WUSB frames. WUSB host also maintains a WUSB cluster address (range between 0x80 to 0xFE), which serves as broadcast address for the WUSB cluster. Cluster address is selected from private device address range of MAC device addresses. Table 9.1 lists the addresses used in WUSB host.

WUSB Device Addressing: If WUSB device is a self beaconing device then it uses unicast generated device address for transmitting beacon. The device address field in WUSB frames is set according to device's WUSB device state. Details of WUSB device states are described in the following sub-sections. WUSB device address is in the range of private device address (between 0x0000 to 0x00FF). Host assigns WUSB device address to the device. Device address range for different WUSB device state is as listed in Table 9.2.

Table 9.1 WUSB host address range.

Type	Range	Description
Unused	0x00 to 0x7F	Unused for host
Cluster broadcast address	0x80 to 0xFE	Cluster address of the WUSBcluster
Unused	0xFF	Unused for host
Host address	0x100 to 0xFEFF	Host device address
Unused	0xFF00 to 0xFFFF	Unused for host

Table 9.2 WUSB device address range.

Type	Range	Description
Default device address	0x00	This address is default WUSB device address
Authenticated WUSB device address range	0x01 to 0x7F	Host assigns device address from this range when normal enumeration is successful and state is Authenticated state.
Unauthenticated WUSB device address range	0x80 to 0xFE	Host assigns the device address from this range to device in response to DN-connect. This is temporary address during authentication.
Unconnected WUSB device address	0xFF	This address is used when device is in Unconnected state
Unused	0x100 to 0xFFFF	Not used as address of the device

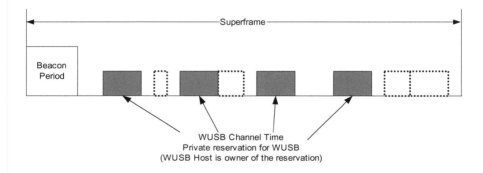

Fig. 9.12 WUSB channel time.

9.3.5 Reservation for Wireless USB Operation

Wireless USB uses UWB radio for actual data communication. MAC has the concept of superframe. WUSB host transmits beacon in beacon period of the superframe, with source address as generated MAC device address. Host reserves MASs with target as private device address (WUSB cluster address) and reservation type as private reservation for WUSB operation. Figure 9.12 shows WUSB channel. Host transmits MMCs in this WUSB channel. WUSB channel time is a free running timer at the host. WUSB devices adapt to the reservation of host. All WUSB control operations and data transfer are done under WUSB channel.

9.3.6 Micro-scheduled Management Command (MMC)

Wireless USB channel is encapsulated in UWB channel by means of DRP reservation. WUSB channel includes sequence of linked frames, which are

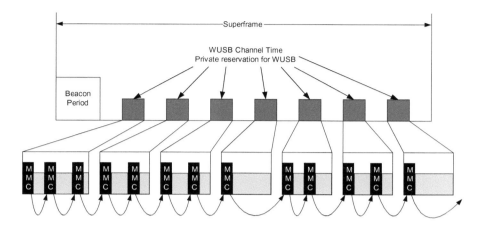

Fig. 9.13 WUSB MMC.

called Micro-scheduled Management Commands (MMC). MMC is transmitted by WUSB host. It is Application Specific Control frame of MAC. Figure 9.13 shows WUSB channel with MMCs. MMC is an engine for WUSB system, which performs management of devices and utilization of WUSB channel time. MMC has different information elements for controlling of transmission and reception of WUSB data.

Basic information contained in MMC is as follows:

- Information about the host.
- Current channel time.
- I/O Control information for all devices.
- Channel time allocation for all devices which serves as a token for the device in USB terminology.
- Timing information for next MMC.
- Information for device notification time slots for asynchronous notifications from device to host.

Wireless USB maintains a free running timer across the WUSB channel. MMC of 24 bit time stamp field is shown in Figure 9.14. Each device synchronizes with the time stamp provided by the host and to act as transmitter or receiver at the time allocated for it.

Microsecond time stamp is of 7 bit, counts from 0 to 124, and wraps to zero after that. One-eighth ms time stamp is of 17 bit, which counts from

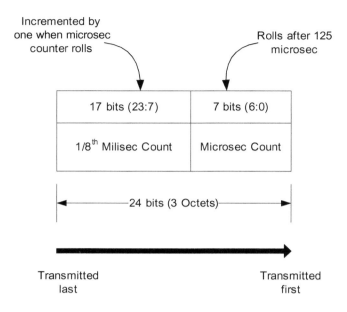

Fig. 9.14 WUSB time stamp.

0 to 131071 (all 17 bits one), and then wraps to zero. This value is incremented by one when microsecond value wraps from 124 to 0.

9.3.7   WUSB Transaction

Wireless USB Transaction consists of the packets transmitted between host and device managed by MMC. Similar to USB, WUSB transaction contains token, data, and handshake, but sequence of packets is different from wired USB. WUSB merges multiple tokens to one in MMC in order to optimize the usage of the wireless medium. Host schedules upstream and downstream together in WUSB channel and reduces the antenna turnaround. Figure 9.15 shows the WUSB transaction in comparison to wired USB transactions.

9.3.8   Device Endpoints and Pipes

Endpoint is a mean of communication flow between host and device. The communication flow which is identified by endpoint is called Pipe. Each WUSB device implements Default control pipe, usually endpoint zero. Pipes

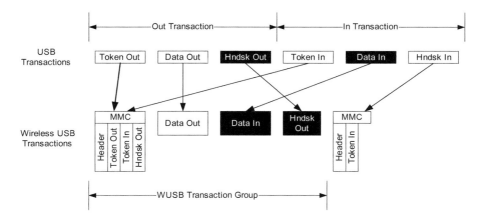

Fig. 9.15 WUSB transaction.

are defined as IN or OUT depending on direction of data to/from the host. Control pipes are bi-directional.

9.3.9 WUSB Transfers

Wireless USB uses different types of transfer depending on the application requirements. These transfers are same as USB transfers.

Bulk Transfer: Bulk is used for transferring large amount of data. It guarantees the delivery of data but not the bandwidth. Bulk is a best effort service. If WUSB channel is relatively free then it provides higher bandwidth for the transfer, whereas if channel is congested then bandwidth is low.

Interrupt Transfer: Interrupt transfer is used for transferring small amount of data with tight service interval requirement. It is used for the data transfer that requires high reliability. Service interval is defined for each interrupt transfer and accordingly bandwidth is reserved on each service interval for interrupt transfer. WUSB supports minimum service interval of 4 ms for interrupt transfer. It provides limited retry if service period is not expired. It also provides guaranteed retry in next service period if multiple transfer failure occurred during a service period.

Isochronous Transfer: It is used for streaming purpose using WUSB. Isochronous transfer supports for error tolerant data transfer with constant rate, and periodic transfers with bounded service interval. It provides guaranteed bandwidth for data transfer with bounded latency, constant data rate through isochronous pipe when data is available. It also provides guaranteed retry if service period is not expired. Additional reliability is provided by adding buffer at the endpoint.

Control Transfer: Each WUSB device is required to implement default control pipe by endpoint zero for control transfer. This transfer is used for device initialization and logical device management. Control pipe provides best effort service, but does not provide any guaranteed bandwidth. Priority of different pending control transfers can be decided by the host. Number of retries for control transfer is also decided by the host.

9.3.10   Notifications

Notification is an asynchronous way for device to communicate with the host. It does not map to any transfers or pipes. Direction of notification is always from device to host. Notifications are sent by device in Device Notification Time Slot (DNTS). DNTS is scheduled by the host periodically as well as whenever required.

9.3.11   WUSB Device States

The device goes through different WUSB device states from the time it starts the operation till stopping of the operation. Figure 9.16 shows different WUSB device states and their operations.

Unconnected State: This is a default state for the device at power on. Device does not have any connection with the host during this time. Disconnection due to timeout, reconnection failure or proper disconnect also leads the device in this state. No data transfer is possible with host during unconnected state. Device address of the device is unconnected device address (0xFF) in this state. Device can only transmit Device notification connect command (DN-connect) to attempt a connection with the host during this state. When host sends responses to DN-connect by connect acknowledgment, device sets the address in unauthenticated device address range (0x80 to 0xFE). This event is device's transition to connected general state, and unauthenticated sub-state.

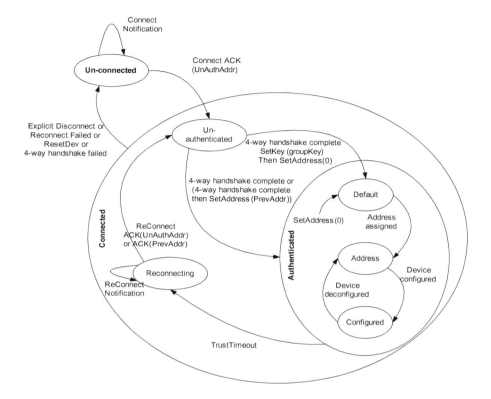

Fig. 9.16 WUSB state machine.

Unauthenticated State: This state is unauthenticated state so most of the communication in this state is not secured. Communication for authentication is done through default control pipe during this state. Device performs 4-way handshake and establishes the PTK. GTK is distributed using PTK. Finally, host sends SetAddress (0) request secured by GTK and device acknowledges it. This is device's transition to authenticated state. If any of these operations are unsuccessful in trust timeout, then device goes back to unconnected state.

Authenticated State: Authenticated State is normal operating state. Sub-states of Authenticated state are same as USB device states. Host assigns the address to the device (from 0x01 to 0x7F) and transits to addressed sub-state from the default sub-state. Host configures the device through different commands, and the device transits to configured state. Host can re-authenticate the device in this state. Normal data communication is done during configured state. Device leaves this state in case of disconnect event, re-authentication failure, or trust timeout. In case of trust timeout, device goes to reconnecting state, otherwise goes to unconnected state.

Reconnecting State: Device tries to reconnect to the host by sending DN-connect during this state. Device address remains unchanged as per earlier state. If host sends the acknowledgment then device enters unauthenticated state, otherwise unconnected state.

9.3.12 WUSB Security and Association

Wireless USB uses the security mechanism provided by MAC. At connection level security between the host and device ensures that device is associated and authenticated and only desired device is communicating with the host. WUSB uses 4-way handshake, generates PTKs and devices get the GTK from the host. Devices are not allowed to communicate to the host without authentication and association.

USB is well known for being very user friendly technology. Due to wired connection, it is very easy to connect the device to a particular host. One of the primary objectives of WUSB is also to provide similar user friendly experience. Various association models are defined and proposed for WUSB, where it can provide good user experience for association of the device (e.g. very simple device with no keypad or screen for selection of a particular host) to a particular host.

9.3.13 WUSB Power Management

Wireless USB defines power management for host and device. WUSB and underlying MAC both are timeslot based. Host and device both know when to be awake for transmission and reception.

Wireless USB device can indicate its intention to go to power save to the host, still be in connected state.

Host can perform power save by sleeping during the time when no traffic is scheduled or expected. It can even increase the duration between two consecutive MMCs to achieve that. Device does not require anything special for this power save other than following instructions through MMCs.

Another way for saving power by host is based on external events like hibernation of platform or radio switch off. It is communicated to device by the host. Devices can even ask the host to wake up by remote wakeup mechanism.

9.4   Implementation Considerations

Wireless USB and ECMA-368 provide basic mechanisms for channel access and interoperation. Enough freedom is also provided to system implementers to only implement minimum (or subset of) required functionalities, so that the host and device can interoperate and perform the functionalities which are required for the product. Innovative and differentiated implementation can be achieved through different application specific options provided in standards, besides methods and mechanisms that are not part of standards.

Following subsections discuss few implementation considerations for WUSB systems.

9.4.1   Device Implementation Considerations

9.4.1.1   Type of Devices

Implementation of WUSB device can vary based on requirement of application or use case. It can support multiple types of devices which can be configured on requirement.

Self beaconing is one of the most desirable types of device implementation which requires almost complete beacon protocol support.

While directed beaconing device does not require beacon protocol or MAC implementation, but it requires implementation of capture packet and transmit packet functions. Transmit and capture functionality is used for transmitting and capturing the beacons in neighborhood of the device. Another consideration for implementation of directed beaconing device is availability of host with support of directed beaconing. As directed beaconing consumes lots of channel bandwidth and support of directed beaconing is very complex for the host, it is possible that very few host vendors support directed beaconing devices. These constraints make directed beaconing less popular type of device implementation.

Nonbeaconing device does not require any MAC functionality, but range is half for Nonbeaconing device. So, it is only suitable for the device which is residing in a very close proximity of the host.

9.4.1.2 Selection of Host

More than one host can be in radio range of a device. To start the operation by sending DN-connect to the host, device has to transmit the beacon and accept the DRP reservation of desired host. Device has to select a desired host and desired DRP reservation, if multiple DRP reservations exist.

DRP reservation is identified by source address of the owner device, target address of the reservation, reservation type (i.e. private), and stream index. While WUSB cluster is identified by host address and WUSB cluster address. For WUSB host, device's source address is WUSB host address, host is always the owner of DRP reservation, DRP reservation target address is WUSB cluster address, and stream index is available in DRP IE. WUSB device can select correct host, join host's beacon group, and join desired WUSB DRP reservation, using these parameters. These parameters are also informed through Host Information IE in MMC.

9.4.1.3 DRP Support for WUSB Device

If UWB device is only acting as WUSB device then generally it connects to only one host at a time, so it is required to support only one DRP reservation (could be represented by more than one DRP IE). Also, WUSB device is always the target of the reservation so it is never required to initiate any reservation.

9.4.2 Host Implementation Considerations

Wireless USB host is a complete UWB device. It is required to support most of the MAC features which are required for smooth operation of the host and good interoperation with non-WUSB UWB devices.

Some of the features of MAC are achieved by WUSB in some other way, e.g. power save is defined by MAC and WUSB both, so WUSB host is required to implement only the power save defined by WUSB, if it is acting only as WUSB host.

Some of the other considerations for host implementation are described briefly below.

9.4.2.1 DRP Support for WUSB Host

Wireless USB host is required to implement DRP reservation of type private only. Best service interval that can be achieved is 4 ms (mainly required for interrupt transfers). The host has to reserve at least one MAS in each zone for that. Preferable reservation for WUSB host is a row reservation to support 4 ms service interval. It may use column reservation for the transfer not requiring tight service interval and requiring continuous time for data transfer (e.g. bulk transfer).

9.4.2.2 Support for Directed Beaconing Device

If WUSB host wants to support directed beaconing WUSB device then it is also required to support for capture packet and transmit packet functions. Also host has to implement the beacon protocol for the devices and generate beacons for the directed beaconing devices from the beacons provided by those devices. These requirements make the host's implementation quite complex and bulky.

9.4.2.3 Host Scheduler

Performance of WUSB host depends on scheduling of the MMCs. The scheduler can be designed depending on requirement and availability of the MASs, number of connected devices, their active endpoints, current traffic requirements, etc. Scheduler is heart of WUSB host and WUSB cluster. Good scheduler can provide best throughput and good user experience to connected devices, while bad scheduler design can make throughput or user experience awful.

9.5 Summary

This chapter explained the usage of UWB common radio platform as WUSB host and device. Here, we have considered different variants of WUSB host and device implementation like self beaconing device, nonbeaconing device, directed beaconing device, host wire adapter, device wire adapter, and dual role device.

This chapter also provides information about the capabilities of UWB common radio platform and how WUSB can use it for channel access and WUSB protocol implementation over DRP reservation. Also, the subsections explain the medium access and scheduling of WUSB host system and device states during the operations.

Implementation considerations during WUSB host and device system design are covered. It explains the usage of UWB common radio and its components required for WUSB systems.

10

Converging Marketplace

10.1 Introduction

Ultra Wideband comes with the promise of catering multiple different applications of various marketplaces such as PC, consumer electronics, and mobile industry. It can replace many existing wired connectivity to wireless and creates opportunity for wireless home and office. UWB is a connectivity technology, which enables very high data rate at short range and without any networking infrastructure. UWB common radio platform is scalable and extendable, so it can be used for variety of applications and protocol adaptations. The radio can also serve for multiple protocols and applications simultaneously. In addition, UWB supports excellent mobility in terms of change of topology. Mobility with very short range and very high data rate makes it suitable for mobile handheld devices.

From industry point of view, UWB has arrived at a very fast pace. Standards of UWB are widely accepted all over the world through different organizations like ECMA, ISO, and ETSI. Regulatory approvals of UWB are also pursued by different alliances and companies quite seriously. UWB common radio is considered by many other industry alliances as its wireless replacement of wired connectivity or as high speed wireless replacement of existing connectivity. Extendable nature of common radio makes it really quick in adaptation to new requirements.

These reasons make UWB as the best choice for high speed short range wireless connectivity. Many expect UWB's fast adoption leading to proliferation of products in the market.

Alternate technologies like IEEE 802.11 is also going high speed in its new avatar 802.11n. Bluetooth, a quite popular WPAN technology originally defined for low data rate applications, is also looking for high data rate for extension of application base with emerging applications.

Ever increasing application requirements, different types of wireless and wired connectivity, high mobility requirements, etc. require continuous changes in technologies. Different network environments and uninterrupted service are the key drivers for new technology development and also a driving factor for existing technologies. It is expected that UWB would be a part of heterogeneous network environment which can provide different kind of services and continuous connectivity.

10.2   Discussion on UWB Products

First Ultra Wideband implementations available in market are Wireless USB products. Many of these are certified by Wireless USB Promoters Group. There are some Certified Wireless USB end products present in market. Other set of UWB implementations are in the form of silicon building blocks from different silicon vendors. One can soon expect wireless internet protocol based UWB products which can be used for internet connectivity, or audio and video transmissions. Efforts are going on for Bluetooth over Ultra Wideband (BToUWB) product development with shorter time to market goals. UWB common radio provides extendibility; hence many companies implement their own protocol adaptation layers for their specific applications. Such vendor specific PALs can provide optimized services for video and audio streaming. Chip vendors have silicon available for specific needs for particular PALs or have common radio to serve different PALs and applications.

Other than consumer products, there is another set of vendors providing UWB test and measurement products. These products include UWB waveform analyzers, protocol analyzers, and protocol analyzers for different PALs (e.g. WUSB and WLP).

10.2.1   Consumer Products

Universal Serial Bus is a very well adopted wired connectivity technology in PC market. Now, it is used in many consumer electronics devices also. UWB becomes the vehicle for unwiring USB through Wireless USB. First Certified Wireless USB products are in the form of dongles, which makes wired USB wireless. These dongles are either host wire adapter or device wire adapter. Some of the native host and device implementations also started entering in the market. Companies like Alereon, Artimi, Intel, NEC Electronics, Realtek

Semiconductors, NXP Semiconductors, LucidPort, Staccato, WiQuest, and Wisair are the leaders to provide the silicon for Wireless USB. Many end product development companies used these silicon blocks and provided end product of HWA and DWA in the market. Companies like Belkin, D-Link, and IOGEAR had used UWB silicon for their HWA and Wireless USB hub (DWA) products. Figures 10.1 and 10.2 show pictures of Wireless USB products from Belkin.

Not being really native wireless system and dependent on the wired interface of USB, throughput of the system depends on wired USB and bulkiness of the protocol stack and drivers. Another class of WUSB products is native

Fig. 10.1 Belkin WUSB HWA.

Fig. 10.2 Belkin WUSB Hub (DWA).

WUSB device and host. These systems generally have various interfaces like PCI, MiniPCI, PCMCI etc. with host PC. Many silicon vendors even provided silicon building blocks and demonstrated better performance compared to dongles with their WUSB systems.

Companies like Dell and Lenovo have adopted WUSB for their laptop computers. These laptops have WUSB host inbuilt and any WUSB device or DWA can connect to that host.

Other than Wireless USB products, semiconductor companies working on UWB are ready with UWB common radio platforms for many different upper layer protocols. Being very high speed communication system, UWB is the right choice for streaming solutions. Companies like Tzero Technologies and Focus Enhancements used UWB for video transfer between different audio–video devices like HDTVs, set-top boxes, video recorders etc.

Another application expected to grow very fast is wireless internet protocol, a solution for *ad-hoc* network and internet infrastructure.

Ultra Wideband common radio also provides the tool for developing vendor specific PALs, so there is very high possibility to have products with vendor specific PALs for various futuristic applications of UWB.

Ultra Wideband is a growing technology so many companies provide their reference design and development kits, which helps in developing new applications, PALs and finally new products.

10.2.2   Test and Measurement Products

With the growing market base of UWB consumer products, test and measurement instruments are required for the development of UWB silicon and consumer products. Numbers of companies are working on UWB products. Test and measurement products help chip vendors, IP providers, firmware and software providers, driver providers and end product developers for validation, performance measurement and standard compliance of the products. Eventually test and measurement products enable faster market reach for the products which are well tested and well performing.

Lecroy and Ellisys are the first companies to provide protocol analyzers for UWB. It supports analysis of ECMA-368 based UWB MAC frames and few PALs.

Figure 10.3 shows Ellisys analyzer box. Ellisys tool includes graphic based environment with run time frame viewing facility which reduces the analysis time drastically. It has capabilities to generate timing diagrams which makes protocol analysis really quick and understandable. The tool features the capability of spooling data to host PC memory without limiting the trace size on internal analyzer memory. Additionally the box supports frame generation tools which can be used for generating various complex test scenarios of the MAC protocol with precise timing requirements via a powerful scripting language. The analyzer and the generator are combined to provide WiMedia platform pre-compliance test facilities. In addition to MAC, the tools also support PAL protocols like wireless USB and Bluetooth over UWB.

Figure 10.4 shows UWB analyzer from Lecroy. It provides graphical user interface with hierarchical frame display in real time. This protocol analyzer is capable of understanding many PAL frame formats and helps debugging and testing of the PAL implementations. It supports multiple UWB radio systems

Fig. 10.3 Ellisys UWB analyzer.

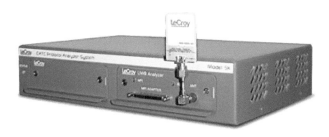

Fig. 10.4 Lecroy UWB tracer.

for testing. It also has capability of frame generation and capture in single unit. Scripting language support can help to generate the scenarios quickly and can be used to perform scenario tests repeatedly and quickly. Lecroy analyzers have capability to capture and generate frames on air or through RF cables. It can also capture frames at MAC–PHY interface level.

Tektronix has developed oscilloscopes for UWB radio. The oscilloscopes provide facility of test, measurement and monitoring of UWB radio based on MB-OFDM technology. These oscilloscopes can be used for analysis of UWB transmitter that includes analysis of RF waveform, demodulation technique, time frequency codes, and data rates. For receiver testing, it can generate UWB waveforms for all time frequency codes and data rates. It also has facility to introduce interferences for realistic testing. These are very useful features for the engineers working on UWB radios based on MB-OFDM, or someone using UWB radio for UWB platform integration and end product development.

For UWB radio test and measurement, Agilent also provides analysis tool set for demodulation analysis and vector-signal analysis. They also provide simulation software for UWB transmitters and receivers. UWB signal generation facility is available with these tools which can be used for UWB receiver testing.

Earlier, analyzers were only available for band group 1, but now there are PHY implementations supporting higher band groups so analyzers are also upgraded with higher band group support. Some of them support multiple radios for scanning multiple channels simultaneously. These tools are expected to be more sophisticated and easy to use along with more features. Tools are improving day by day with advanced features and helping developers for easy testing and better test coverage for improved performance of the end products.

10.3   Bluetooth over UWB: Interesting Paradigm

Bluetooth has been around as WPAN technology leader since long. It is a short range wireless technology with operation range around 10 meters. Bluetooth was designed for low bandwidth wireless connections up to 1 Mbps. It provides proven wireless connectivity with features like low power, low cost, *ad-hoc* networking, and security. It is designed for mobile handhelds with very user friendly association methods. Today, Bluetooth is used in cell phones, cordless phones, hands-free, computers, computer peripherals (mouse, keyboards, etc.), LAN, PDAs, digital cameras etc. In fact, Bluetooth has become so popular that it is difficult to imagine a cell phone without Bluetooth support.

Traditionally, Bluetooth has been very low throughput system compared to UWB and could not be used in streaming media other than audio streaming. Bluetooth over UWB is an evolving paradigm which enables high speed wireless while maintaining very low power. Bluetooth SIG has selected MB-OFDM UWB radio based on ECMA-368 for high data rate wireless Bluetooth system. This new version of Bluetooth technology is expected to meet the high speed demands of synchronizing and transferring large amounts of data as well as enabling high quality video and audio applications for portable devices, multimedia projectors, and television sets. At the same time, Bluetooth will continue catering low speed and low power applications like mouse, keyboard, and headsets.

Bluetooth over UWB architecture is dual radio system, which allows maximizing and leveraging benefits of both Bluetooth and UWB in one system. Bluetooth is very mature in its existing protocol for connection, maintenance, and controlling. These functionalities including low speed data transfers are performed through Bluetooth radio. UWB radio is used as it is good in high speed and high volume data transfer. System is expected to be capable of switching the radio depending on requirement of the applications.

In order to build on the success of Bluetooth in the marketplace, it is very critical that the UWB technology be compatible with Bluetooth and maintains the core and proven attributes. Low power, low cost, *ad-hoc* networking, built-in security features, and ability to integrate into mobile devices are proven features of Bluetooth. UWB comes with all those including high data rate. Dual radio architecture keeps all features of Bluetooth, and provides backward compatibility with the millions of Bluetooth devices currently present in the market.

The Bluetooth SIG has adopted MB-OFDM UWB technology as alternate MAC and PHY, which is capable of meeting the requirements for high speed Bluetooth. The two organizations (Bluetooth SIG and WiMedia Alliance) has committed to work together to ensure that the combined high speed solution is optimized for mobile devices with very low power consumption. Groups are working on creating a draft specification for protocol adaptation layer for

Bluetooth adaptation. Market is ready to see the Bluetooth technology/UWB solution chip sets soon, which can make high speed Bluetooth a reality.

10.4 UWB and IEEE 802.11n

As wireless multimedia is becoming very integral part of consumer electronics devices and PC systems, support for high data rate is becoming very important for all wireless standards. Not being late in race, IEEE 802.11 is also working on the same direction via 802.11n, so it can provide wireless multimedia, transfer of video, HDTV streaming, etc. 802.11n is trying to solve the challenge of high data rate by the extension of existing WLAN. Though both UWB and 802.11n are on separate development paths and by different groups, but in the end, could be addressing the same design requirements. Thus, they are raising the question, whether they are competitors or compliments?

IEEE 802.11n builds on previous 802.11 standards by adding Multiple Input Multiple Output (MIMO) and 40 MHz operation to PHY. MIMO provides transmitter and receiver diversity; it improves the system performance by data rate enhancement. 802.11n offers a long range of up to 200 meters and data rate up to 100 Mbps. It makes the WLAN technology a rival of very popular wired fast Ethernet. MIMO provides powerful and high data rate system which is yet cost effective, therefore it is viewed as the most likely contender for the home network backbone also. On the other hand, UWB is designed to make best use of its low power, high speed operations in short range equipment interconnect such as Personal Computers (PC) and portable equipment including home networking.

IEEE 802.11n has provided very good radio range up to 200 meters. While UWB provides range up to 10 meters with highest power and up to 3 meters at highest data rate. This range of 802.11n is a result of MIMO which includes complexity of implementation and involves increase in power consumption.

IEEE 802.11n provides throughput from 20 Mbps to 100 Mbps, while MB-OFDM based UWB provides PHY data rate up to 480 Mbps and throughput depending on the type of application or PAL varies from 80 Mbps to more than 300 Mbps. Traditionally 802.11 MAC was designed for internet protocol and bursty traffic, but UWB MAC based on ECMA-368 is capable of bursty traffic and streaming traffic with tight service interval requirement. Capability to support multiple PALs on single radio and multiple types of applications render unique proposition to UWB system for different WPAN environment.

To be successful in the market, next generation connectivity system must be optimized for low power consumption. Power consumption is one of the most important requirements for WPAN devices being portable, mobile, and battery-operated. IEEE 802.11n is relatively long range LAN system so it is more optimized for long range, on other hand UWB which is short range is optimized for low power with short range. UWB also supports transmit power control and maintains the link with minimum power required, which is a regulatory requirement for this technology.

Next generation mobile devices and laptops are expected to have different kinds of radio on same system. These different radios are prone to interference from each other. Other radios present in wireless ecosystem can also interfere with each other. Current 802.11 systems operate in the 2.4 GHz or 5 GHz range. This spectrum is also used by numerous other wireless technologies, including Bluetooth, WiMAX, cordless telephones, and microwave ovens. Commercial UWB system operating in 3.1 to 10 GHz has comparatively less interference from other radios. Power levels of these systems are regulated to be very low, so it does not create realistic interference to other systems. Detect and Avoid mechanism, as discussed in Chapter 4, is employed in UWB system to avoid any possible interference to other communication technology users.

MB-OFDM UWB system specification is developed by WiMedia Alliance and adopted by ECMA and ISO standard. Further, the certification processes for PHY and radio platforms are evolved. Prototype systems and initial products like Wireless USB, USB based HWA and DWA, and wireless streaming solutions based on UWB radio are available. This lead time for UWB augurs well for future of UWB based products.

Ultra Wideband radio is suitable for WPAN with low power consumption and high throughput. System which requires comparatively long range, which is really a WLAN environment, can use 802.11n system with high throughput. For different high speed streaming requirement of home environment, PC peripheral and mobile power sensitive device UWB radio is a better choice in terms of performance at short range with less power.

10.5 Convergence of Technologies and Devices

We are living in era of communication where information exchange is a must in our life. Information could be text, voice, audio, video, high quality video, 3D, or mix of many of these things. Today access of information is possible through many different mediums such as network cables, cables for wireline phone, satellites, base stations for 3G, WiFi, Bluetooth etc. Still these networks are not interconnected or not fully interconnected. One could foresee next generation network as network ecosystem of interconnected and interworked devices that can be accessed through different mediums and can be reached from any where.

Next generation networks are expected to be packet based networks that can provide services including telecommunication services and can make use of multiple broadband and QoS enabled transport technologies. Service related functions are independent from underlying transport technologies. It can support generalized mobility and allow consistent service to user without caring about who is the service provider of the network. Users may access the network through different technologies, may converge through different technology domains to access the information. Depending on situation user remains connected in the best possible way. Growth of IP based network and services leads to great improvement in wired and wireless network backbones so it can serve for required bandwidth and QoS. User can access data as well as voice service with same network technology.

We are getting into an era where services for end user are portable. Device is able to access the service either being in packet switch network of 3G, 4G, circuit switched network or any other network. Again, because of converging network technologies, short range wireless technologies can interwork with wide area network (wired and wireless). Mesh networking technologies using short range wireless can be part of heterogeneous next generation network. One of the examples is WLAN convergence with 2.5G and 3G, where end user (WLAN station) has capability to access WLAN services and 3G services both. Similarly, there are possibilities of WPAN integrations and interworking with wide area network and back bones to get those services through WPAN. UWB WPAN can provide the mesh network infrastructure and back bone connectivity through different PALs. Single radio and multiple application architecture and combo chips with multiple different radios make the device capable of communicating with different network technologies and different application layer protocols.

Such portability allows device to move seamlessly from macro area (2.5G, 3G or 4G network) to micro area (WPAN or WLAN network) and access wide range of services of all networks. Further, it allows creating the *ad-hoc* network

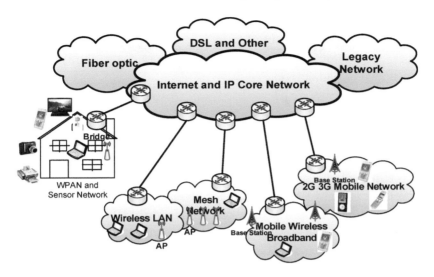

Fig. 10.5 Heterogeneous network.

for the services they want locally and also can connect to peripherals wirelessly or create body area network, which also can converge to heterogeneous network. (see Figure 10.5) These convergent technology enabled devices can seamlessly move across any kind of networks with unified services.

Next generation network requires new SoCs for end devices and different network elements. Similar to any other system, it is required to be optimal in power, performance, size, battery life etc. Additionally, the converging architecture includes multiple components like multiple radios, multiple connectivity stacks, and multiple wired and wireless connections, bridging components, so complexity is really high compared to traditional dedicated systems. Further, applications over those SoCs are different than usual. The system design of these evolving technologies is expected to improve with the changing and dynamic market conditions.

Will the packet based network and convergence through different network affect the performance and throughput? Also can the service which is accessed through one kind of network provide same performance when accessed through different kind of network? Answers to these questions will decide the future of next generation network. Device manufacturers, service providers, semiconductor developers will have to envision the heterogeneity in terms of formation of various macro, micro and *ad-hoc* networks, then allow for seamless movement of users across these networks. Applications, services, and features need to be defined to ensure compatibility across networks, domains, and devices.

10.6 Summary

This chapter provides the market overview of UWB products. It discussed various products available in marketplace, classifications of those and what more is expected in near future.

UWB adoption by Bluetooth SIG for Bluetooth's requirement of high data rate is creating exciting prospects for early adoption of UWB in variety of markets. We also discussed 802.11n which is being argued as one of technologies to watch, in the context and background of UWB developments.

Ultra Wideband unwires many existing applications and identifies new applications and provides great opportunity to replace the cables. UWB has to interwork with other technologies for wide adoption. Future applications will continue to expect even higher data rates. Similarly, other competing technologies will also continue their development. In this situation, UWB developers must come up with better products for success of the technology and complete utilization of the capabilities of UWB for leading the market.

Abbreviations

2G	2nd Generation
3G	3rd Generation
4G	4th Generation
3GPP	3rd Generation Partnership Project
3GPP2	3rd Generation Partnership Project 2
AC	Access Category
AC_BE	Access Category–Best Effort
AC_BK	Access Category–Background
ACK	Acknowledgement
AC_VI	Access Category–Video
AC_VO	Access Category–Voice
ADC	Analog to Digital Converter
AES	Advanced Encryption Standard
AIFS	Arbitration Inter-Frame Space
ANSI	American National Standards Institute
AS	Application Specific
AS IE	Application Specific IE
AWGN	Additive White Gaussian Noise
B-ACK	Block-ACK
BG	Beacon Group
BOK	Bi-orthogonal Keying
BP	Beacon Period
BPO IE	Beacon Period Occupancy IE
BPSK	Binary Phase Shift Keying
BPST	Beacon Period Start Time
BToUWB	Bluetooth over UWB
BT SIG	Bluetooth Special Interest Group

CBC-MAC	Cipher Block Chaining — Message Authentication Code
CCA	Clear Channel Assessment
CCM	Counter Mode Encryption and Cipher Block Chaining Message Authentication Code
CDMA	Code Division Multiple Access
CE	Consumer Electronics
CMOS	Complementary Metal-oxide Semiconductor
CSMA	Carrier Sense Multiple Access
CSMA/CA	Carrier Sense Multiple Access — Collision Avoidance
CSM	Common Signaling Mode
CRC	Cyclic Redundancy Check
CTS	Clear to Send
CW	Contention Window
DAA	Detect and Avoid
DAC	Digital to Analog Converter
DCM	Dual-Carrier Modulation
DevAddr	Device Address
DN	Device Notification
DP	Data Period
DRD	Dual Role Device
DRP	Distributed Reservation Protocol
DRP IE	Distributed Reservation Protocol IE
DSP	Digital Signal Processor
DS-SS	Direct Sequence Spread-Spectrum
DS-UWB	Direct Sequence-Ultra Wideband
DWA	Device Wire Adapter
DVD	Digital Video Disc
DVR	Digital Video Recorder
DWA	Device Wire Adapter
ECC	Electronic Communications Committee
ECMA	European Computers Manufacturers Association
EIRP	Effective Isotropic Radiated Power
EO	Encryption Offset
ERO	European Radio Communication Office
ETSI	European Telecommunications Standards Institute
EUI	Extended Unique Identifier

EV-DO	Evolution-Data Optimized
EV-DV	Evolution-Data/Voice
FCC	Federal Communications Commission
FCS	Frame Check Sequence
FEC	Forward Error Correction
FEQ	Frequency Domain Equalizer
FFI	Fixed Frequency Interleaving
FFT	Fast Fourier Transform
GPRS	General Packet Radio Service
GPS	Global Positioning System
GSM	Global System for Mobile Communications
GSSI	Geographical Survey Systems, Inc
GTK	Group Temporal Key
GTKID	Group Temporal Key Identifier
HCS	Header Check Sequence
HDTV	High Definition Television
HWA	Host Wire Adapter
IC	Integrated Circuit
ID	Identifier
IDFT	Inverse Discrete Fourier Transform
IE	Information Element
IEEE	Institute of Electrical and Electronics Engineers
IFFT	Inverse Fast Fourier Transform
Imm-ACK	Immediate-ACK
IPv4	Internet Protocol version 4
IPv6	Internet Protocol version 6
IrDA	Infrared Data Association
ISO	International Organization for Standardization
KCK	Key Confirmation Key
L2CAP	Logical Link Control and Adaptation Layer
LAN	Local Area Network
LDC	Low Duty Cycle
LMP	Link Manager Protocol
LNA	Low Noise Amplifier
LTE	Long Term Evolution

MAB IE	Multicast Address Binding IE
MAC	Medium Access Control
MAN	Metropolitan Area Network
MAS	Medium Access Slot
MBOA	Multi-band OFDM Alliance
MB-OFDM	Multi-band Orthogonal Frequency Division Multiplexing
MCDU	MAC Command Data Unit
MIC	Message Integrity Code
MIFS	Minimum Inter Frame Space
MIMO	Multiple Input Multiple Output
MiniPCI	Mini Peripheral Component Interconnect
MKID	Master Key Identifier
MKID IE	Master Key Identifier IE
MMC	Micro-scheduled Management Command
MPDU	MAC Protocol Data Unit
MSDU	MAC Service Data Unit
MUX	Multiplexing
NAV	Network Allocation Vector
NFC	Near Field Communication
ODM	Original Design Manufacturer
OEM	Original Equipment Manufacturer
OFDM	Orthogonal Frequency Division Multiplexing
OS	Operating System
OUI	Organizationally Unique Identifier
PAL	Protocol Adaptation Layer
PAN	Personal Area Network
PAR	Project Authorization Request
PC	Personal Computer
PCA	Prioritized Contention Access
PCI	Peripheral Component Interconnect
PCMCI	Personal Computer Memory Card International
PDA	Personal Digital Assistant
PDC	Personal Digital Cellular
PER	Packet Error Rate
PHY	Physical Layer
PLCP	Physical Layer Convergence Protocol

PLL	Phase Locked Loop
PLME	Physical Layer Management Entity
PMD	Physical Medium Dependent
PPDU	PLCP Protocol Data Unit
ppm	Parts Per Million
PSDU	PLCP Service Data Units
PTK	Pair-wise Temporal Key
QoS	Quality of Service
QPSK	Quadrature Phase Shift Keying
RFID	Radio Frequency Identification
RM	Range Measurement
RMR	Range Measurement Report
RRQ	Range Measurement Request
RSSI	Receive Signal Strength Indicator
RTS	Request to Send
SDP	Service Discovery Protocol
SFC	Secure Frame Counter
SFN	Secure Frame Number
SIFS	Short Inter Frame Space
SIG	Special Interest Group
SNR	Signal to Noise Ratio
SoC	System-on-Chip
TCP	Transmission Control Protocol
TDMA	Time Division Multiple Access
TFC	Time Frequency Code
TFI	Time Frequency Interleaving
TFI2	Two-band Time Frequency Interleaving
TG	Task Group
TIA	Telecommunications Industry Association
TIM IE	Traffic Indication Map IE
TKID	Temporal Key Identifier
TPC	Transmit Power Control
TV	Television
TWTT	Two Way Time Transfer
UDA	Unused DRP Reservation Announcement
UDR	Unused DRP Reservation Response

USB	Universal Serial Bus
USB-IF	Universal Serial Bus Implementers Forum
UTM	Unified Threat Management
UWB	Ultra Wideband
VPN	Virtual Private Network
WAN	Wide Area Network
WBAN	Wireless Body Area Network
W-CDMA	Wideband Code Division Multiple Access
WHCI	WUSB Host Controller Interface
WiMAX	Worldwide Interoperability for Microwave Access
WiMedia	WiMedia Alliance
WLAN	Wireless Local Area Network
WLP	WiMedia Link Protocol
WMAN	Wireless Metropolitan Area Network
WPAN	Wireless Personal Area Network
WSN	Wireless Sensor Network
WUSB	Wireless Universal Serial Bus
WWAN	Wireless Wide Area Network
ZPS	Zero-Padded Suffix

References

[1] 3GPP, <www.3gpp.org>.

[2] 3GPP2, <www.3gpp2.org>.

[3] P. Belanger and Ken Biba, *802.11n Delivers Better Range*, Online, May 31, 2007. <http://www.wi-fiplanet.com/tutorials/article.php/3680781>.

[4] 802.1D — *Expedited Traffic Network Protocols and Standards*, Online, Sep 23, 2004. <http://suraj.lums.edu.pk/~cs573w06/lectures/06_expedited_traffic.ppt>.

[5] Dr. R. Fontana, "Multispectral solutions." *A Brief History of UWB Communications*, Online, May 17, 2004. <http://www.ferret.com.au/n/A-brief-history-of-UWB-communications-n681131>.

[6] J. Salokannel and J. Reunamaki, Nokia, "Adaptive beacon period in a distributed network." International Application No.: PCT/IB2005/001162, Publication Date: Nov 10,2005, International Filing Date: March 29, 2005. <http://www.wipo.int/pctdb/en/wo.jsp?IA=IB2005001162&DISPLAY=STATUS>.

[7] Agilent UWB tools, Online <http://www.home.agilent.com/agilent/industry.jspx?nid=-34909.0.00&lc=eng&cc=US>.

[8] P. Seebach, *A House Divided, UWB's Double Standards*, Online, Jan 25, 2005, <http://www.ibm.com/developerworks/library/pa-spec4.html>.

[9] Alereon, <http://www.alereon.com>.

[10] D. Blankenbeckler, *An Introduction to Bluetooth.* <http://www.wirelessdevnet.com/channels/bluetooth/features/bluetooth.html>.

[11] Dr. W. Yeh. Keming and Dr. L. Wang, *An Introduction to the IrDA Standard and System Implementation*, Online <http://www.actisys.com/article.html>.

[12] N. Hunn, "An introduction to Wibree." An EZURiO white paper, Online, 2006. <http://www.tdksys.com/files/00616.pdf>.

[13] R. Kolic, *An introduction to Wireless USB (WUSB)*, Online, Feb 21, 2004. <http://www.deviceforge.com/articles/AT9015145687.html>.

[14] ANSI, <www.ansi.org>.

[15] R. Aiello, J. Ellis, and L. Taylor, "General atomics." *Application Opportunities for High Rate WPANs*, IEEE 802.15.02/143, Online, 11 March 2002. <http://grouper.ieee.org/groups/802/15/pub/2002/Mar02/02143r0P802-15_SG3a-Application-Opportunities-GA.ppt>.

[16] Artimi, <http://artimi.com>.

[17] S. Jogi, P. Wason, B. Holur, M. Choudhary, W. Yoon, and Y. Kim, "Samsung." "Associated devices information element and faster association." *IEEE 802.15-04-0144-00-003b*, Online, March 17, 2004. <https://mentor.ieee.org/802.15/

file/04/15-04-0144-00-003b-15-04-0144-00-003b-associated-devices-information-element-and-faster-association.ppt>

[18] J. Geier, *A Technology to Consider: Ultrawideband*, Online, February 25, 2003. <http://www.wi-fiplanet.com/tutorials/article.php/1598581>.

[19] A. Lewicki, J. del Prado Pavón, J. Talayssat, E. Dekneuvel, and G. Jacquemod, "A virtual prototype for bluetooth over ultra wide band system level design." *IEEE Explore, Design Automation and Test in Europe, 2008*, Online, March 2008. <http://www.ieeexplore.ieee.org/xpls/abs_all.jsp? isnumber=4484624&arnumber= 4484916&count=312&index=291>.

[20] E. Jovanov, A. Milenkovic, C. Otto, and P. C de Groen, "A wireless body area network of intelligent motion sensors for computer assisted physical rehabilitation." *Journal on Neuroengineering Rehabilitation*, Vol. 2, no. 6, 2005. Published online, March 1, 2005. <http://www.pubmedcentral.nih.gov/articlerender.fcgi?artid=552302>.

[21] *Bluetooth SIG Selects WiMedia Alliance Ultra-Wideband Technology for High Speed Bluetooth Applications*, Online, Mar 28, 2006. http://www.bluetooth.com/bluetooth/ press/sig/bluetooth_sig_selects_wimedia_alliance_ultrawideband_technology_for_ high_speed_bluetooth_application.htm.

[22] D. Chomienne and M. Eftimakis, "NewLogic." *Bluetooth Tutorial*, Online, 2001. <http://www.cs.umu.se/kurser/TDBD16/VT07/Bluetooth-Tutorial-2001.pdf>.

[23] Consumer Electronics Show, <http://www.cesweb.org>.

[24] K. Lee, J. Park, and K. Moon, "Convergence of high-speed powerline communication and WiMedia UWB for multimedia home networks." *IEEE Explore*, Electronics and Telecommunication Research Institute, Daejeon, R. O. Korea, Online. <http://ieeexplore.ieee.org/iel5/4492759/4510387/04510414.pdf>

[25] C. Razzell, "Distance based association for Ultra Wide Band (UWB) Wireless Personal Area Network (PAN) devices." International Patent Application No: PCT/IB2006/050664, October 08, 2006.

[26] F. daCosta, "MeshDynamics." *Dynamic Beacon Alignment in Simultaneously Operating Piconets (SOP) Using the Heart Beat Approach*, IEEE P802.15-04/135r0, March 16, 2004. <https://mentor.ieee.org/802.15/file/04/15-04-0135-00-0005-dynamic-beacon-alignment-for-sops.doc>.

[27] J. Barr, "DS-UWB PAR comments resolution." *IEEE 802.15-04/0454r1*, Online, Sept, 2004. <https://mentor.ieee.org/802.15/file/04/15-04-0454-01-003a-ds-uwb-par-comment-resolution.ppt>.

[28] ECMA International, <www.ecma-international.org>.

[29] *Ellisys: WiMedia Explorer 300 Analyzer (Wireless USB and UWB Protocol Analyzer)*, Online <http://www.ellisys.com/products/wusbex300/index.php>.

[30] *Ellisys: WiMedia Explorer 300 Generator (WiMedia-based Byte Level Frame Generator)*, Online <http://www.ellisys.com/products/uwbgen320/index.php>.

[31] IEEE, *EtherType Tutorial*, Online <http://standards.ieee.org/regauth/ethertype/ type-tut.html>.

[32] ETSI, <www.etsi.com>.

[33] J. G. Andrews, A. Ghosh, and R. Muhamed, *Fundamentals of WiMAX: Understanding Broadband Wireless Networking*, Prentice Hall PTR, March 9, 2007.

[34] IEEE, *Guidelines for Use of a 48-bit EXtended Unique Identifier (EUI-48)*, Online <http://standards.ieee.org/regauth/oui/tutorials/EUI48.html>.

[35] ECMA-368, *High Rate Ultra Wideband PHY and MAC Standard*. Second Ed., December 2007.

[36] Focus Enhancements, <http://www.focussemi.com>.

[37] T. W. Barrett, *History of UltraWideBand (UWB) Radar & Communications: Pioneers and Innovators, Progress In Electromagnetics Symposium 2000 (PIERS2000)*, Cambridge, MA, Online, July 2000. <www.ntia.doc.gov/osmhome/uwbtestplan/barret_history_(piersw-figs).pdf>.

[38] HomeRF, <http://en.wikipedia.org/wiki/HomeRF>.

[39] B. Vertenten and S. Yuxi, *How to Mitigate Packet Loss in Wireless USB Net-working Applications*, Online, October 07, 2008. <http://www.embedded.com/design/networking/210700326>.

[40] IEEE, <www.ieee.org>.

[41] IEEE 802.11n, <http://en.wikipedia.org/wiki/IEEE_802.11n>.

[42] Infrared Data Association, <http://en.wikipedia.org/wiki/Infrared_Data_Association>.

[43] C. Brabenac, "Intel SG3a CFA response — wireless peripherals." *IEEE 802.15-02/139r0*, Online, March 08,2002. <http://grouper.ieee.org/groups/802/15/pub/2002/Mar02/02139r0P802-15_SG3a-Intel-CFA-Response-Wireless-Peripherals.ppt>.

[44] Intel, *Ultra-Wideband Technology*, <http://www.intel.com/technology/comms/uwb>.

[45] Dr. J. Sharony, *Introduction to Wireless MIMO — Theory and Applications*. November 15, 2006. <http://www.ieee.li/pdf/viewgraphs_wireless_mimo.pdf>.

[46] IrDA, <http://www.irda.org>.

[47] ISO, <www.iso.org>.

[48] B. Holur, "Key drivers to next generation networks." *Times Global Journal* Vol. 1, Online <http://www.sasken.com/downloads/TGJ/issue1/key_drivers-next_gen_networks.html>.

[49] IEEE, *Known EtherType List*, Online <http://standards.ieee.org/regauth/ethertype/eth.txt>.

[50] R. Roberts, K. Siwiak, and J. Ellis, *IEEE P802.15.SG3a PAR, IEEE P802.15-02/372r2*, Online, Nov, 2002. <http://ieee802.org/15/pub/2002/Nov02/02370r2P802-15_SG3a-PAR.doc>.

[51] Lecroy: *UWBTracer / UWBTrainer*, Online <http://www.lecroy.com/tm/products/ProtocolAnalyzers/UltraWideband/UWB_Tracer_Trainer/default.asp?menuid=66>.

[52] LucidPORT, <http://www.lucidport.com>.

[53] ECMA-369, *MAC-PHY Interface for ECMA-368*. Second Ed., December 2007.

[54] J. Ding, L. Zhao, S. R. Medidi, and K. M. Sivalingam, *MAC Protocols for Ultra-Wide-Band (UWB) Wireless Networks: Impact of Channel Acquisition Time*, School of EECS, Washington State University, Online, <http://dawn.cs.umbc.edu/Papers/2002/ITCOM02-UWB.pdf>.

[55] P. Wason, B. S. Holur, M. Choudhary, S. D. Jogi, and T. Arunan, "Method for media access controlling and system and method for channel time reservation in distributed wireless personal area network." United States Patent Application: 20060198353, Online, September 7, 2006. Samsung Electronics Co. Ltd.

[56] MIMO, <http://en.wikipedia.org/wiki/Multiple-input_multiple-output>.

[57] E. Biglieri, R. Calderbank, A. Constantinides, A. Goldsmith, A. Paulraj, and H. Vincent Poor, *MIMO Wireless Communications*, Cambridge University Press, January 8, 2007.

[58] Moore's Law, <http://en.wikipedia.org/wiki/Moore%E2%80%99s_law>.

[59] R. Aiello, T. Larsson, D. Meacham, Y. Kim, and H. Okado, "Multi-Band performance tradeoffs." *IEEE 802.15-03/209r1*, Online, May 2003. <http://ieee802.org/15/pub/2003/May03/03209r1P802-15_TG3a-Multiband-Performance-Tradeoffs.ppt>.

[60] NEC Electronics, *Wireless USB*, <http://www.necel.com/cwusb/en/index.html>.

[61] NFC Forum, <www.nfc-forum.org>.

[62] NXP Semiconductors, *Wireless USB*, <http://www.nxp.com>.

[63] R. Prasad, *OFDM for Wireless Communications Systems*, Artech House Publishers, August 31, 2004.

[64] J. Heiskala and J. Terry, *OFDM Wireless LANs: A Theoretical and Practical Guide*, Sams, December 21, 2001.

[65] Ellis, Siwiak, Roberts, *P802.15. 3a Alt PHY Selection Criteria*, March 2003, <http://grouper.ieee.org/groups/802/15/pub/2003/Jul03/03031r10P802-15_TG3a-PHY-Selection-Criteria.doc>.

[66] T. M. Siep, Ian C. Gifford, M/A-COM, Incorporated, Richard C. Braley, and Robert F. Heile, Paving the way for personal area network standards: An overview of the IEEE P802.15 working group for wireless personal area networks." *IEEE Personal Communications*, Online, Feb 2000. <http://www.comsoc.org/pci/private/2000/feb/Gifford.html>.

[67] PCI SIG, <www.pcisig.com>.

[68] T. Arunan, S. D. Jogi, M. Choudhary, B. S. Holur, P. Wason, and Y.-S. Kim, "Power saving system in distributed wireless personal area network and method thereof." United States Patent Application 20050288069, Samsung Electronics Co. Ltd., Online, December 29, 2005.

[69] J. Decuir, "MCCI," *Progress in Ultra Wideband*, Online <http://www.ee.washington.edu/research/ieee-comm/Presentations/comsoc_talks/UWB4Q05_jdecuir.ppt>.

[70] Radio Frequency Identifier (RFID), <http://en.wikipedia.org/wiki/RFID>.

[71] Realtek, *Ultra-Wideband ICs*, <http://www.realtek.com.tw/products/productsView.aspx?Langid=1&PNid=13&PFid=13&Level=2&Conn=1>.

[72] Standards Make Good Business Sense, Online <http://standards.ieee.org/sa-mem/why_std.html>.

[73] P. Johansson, Chair, 1394 TA Wireless Working Group, *Status Report: IEEE 1394 PAL for WiMedia MAC*, 1394 TA 2nd Quarter Meeting, Seoul, Korea, Online, April 20, 2005. <http://www.1394ta.org/Secure-WWW/Secure-html/wireless/Presentations/0105-r00-20-Apr-05-WiMedia_MAC_PAL_Status.pdf>.

[74] So, Who Needs ZigBee?, Online, Nov 8, 2005. <http://rfdesign.com/next_generation_wireless/who-needs-zigbee/>.

[75] Staccato Communications, <http://www.staccatocommunications.com>.

[76] P. Wason, S. D. Jogi, B. S. Holur, M. Choudhary, and T. Arunan, "System and method for channel time reservation in distributed wireless personal area network." United States Patent Application 20050259617, Samsung Electronics Co. Ltd., November 24, 2005.

[77] R. A. Ughreja, and S. D. Jogi, "System and method for communication between wireless universal serial bus hosts." Indian Patent Application: 1106/CHE/2005, published 2007-09-21, filed on 2005-08-10.

[78] S. Jogi, P. Wason, B. S. Holur, and Dr. M. Choudhary, "System and method for medium access control in wireless mobile ad-hoc networks." Indian Patent Application

Number :221/CHE/2004, Date Of Filing: 12/03/2004, Patent No. 211564, Date Of Grant: 05/11/2007.

[79] E. R. Green and S. Roy, "System architectures for high-rate ultra-wideband communication systems: A review of recent developments." *Intel Labs*, Online <www.intel.com/technology/comms/uwb/download/w241_paper.pdf>.

[80] S. Jogi, P. Wason, M. Choudhary, B. S. Holur, T. Arunan, and Y.-S. Kim, "System for dynamically shifting beacons in distributed wireless network and method thereof." United States Patent Application 20060018298, January 26, 2006.

[81] M. Ilyas, *The Handbook of Ad Hoc Wireless Networks*, CRC, December 20, 2002.

[82] B. O'Hara, and A. Petrick, *The IEEE 802.11 Handbook: A Designer's Companion*, IEEE Press, 1 January 2005.

[83] Y.-H. Tseng, *The MAC Issue for UWB*, National Taiwan University, Online <http://inrg.csie.ntu.edu.tw/2002/The%20MAC%20Issue%20for%20UWB.ppt>.

[84] J. del Prado Pavón, N. Sai Shankar, V. Gaddam, K. Challapali, and C.-T. Chou, "The MBOA-WiMedia specification for ultra wideband distributed networks." *IEEE Explore, Communication Magazine, IEEE*, Vol. 44, no. 6, Online, June 2006. <http://ieeexplore.ieee.org/xpls/abs_all.jsp?arnumber=1668431>.

[85] Y. Solomon, "The role of interoperability in the market for connectivity technologies." WiMedia Alliance White Paper Online, April 6, 2006. <http://www.wimedia.org/imwp/idms/popups/pop_download.asp?contentID=8078>.

[86] J. Decuir, "Two Way Time Transfer based ranging." *IEEE 802.15-04a/0573r0*, Online, October 6, 2004. <https://mentor.ieee.org/802.15/file/04/15-04-0573-00-004a-two-way-time-transfer-based-ranging.ppt>.

[87] The Ultra Wideband Blog, *Sponsored by Tzero*, Online <http://uwbblog.typepad.com/>.

[88] *Tektronix UWB test tools*, Online <http://www.tek.com/Measurement/applications/serial_data/wimedia.html>.

[89] Tzero Technologies, <http://www.tzti.com>.

[90] F. Bhesania and P. G. Madhavan, *UltraWide Band Architecture for Windows*, Microsoft Corporation, Online, <http://download.microsoft.com/download/5/b/9/5b97017b-e28a-4bae-ba48-174cf47d23cd/CON092_WH06.ppt>.

[91] C. Snow, "Ultra wideband communications past, present, and future." *Second UBC-IEEE Workshop on Future Communications System*, Online, Mar 9, 2007. <http://bul.ece.ubc.ca/UWB_UBC.pdf>

[92] E. Griffith, *Ultrawideband Groups Merge*, Online, March 3, 2005. <http://www.wi-fiplanet.com/news/article.php/3487246>.

[93] R. Aiello and A. Batra, eds., *Ultra Wideband Systems, Technologies and Applications*, Newnes, June 12, 2006.

[94] Y. M. Kim, *Ultra Wide Band (UWB) Technology and Applications*, Online, July 10, 2003. <http://www.cse.ohio-state.edu/siefast/presentations/ultra-wide-band-kimyoung-2003/ultra-wide-band-kimyoung-2003.ppt>.

[95] Ultra-Wideband (UWB) *Technology, Enabling High-Speed Wireless Personal Area Networks, Intel Whitepaper*, Online <http://www.intel.com/technology/ultrawideband/downloads/Ultra-Wideband.pdf>.

[96] *Ultra-Wideband (UWB) Technology, One Step Closer to Wireless Freedom*, Online <http://www.intel.com/technology/comms/uwb/download/wireless_pb.pdf>.

[97] *Universal Serial Bus Revision 2.0 Specification*, USB Implementers Forum, Inc., Online, April 2000. <http://www.usb.org/developers/docs/usb_20_040908.zip>.

[98] USB, <www.usb.org>.

[99] IEEE, *Use of the IEEE assigned Organizationally Unique Identifier with ANSI/IEEE Std 802-2001 Local and Metropolitan Area Networks*, Online <http://standards.ieee.org/regauth/oui/tutorials/lanman.html>.

[100] *UWB — Best Choice to Enable WPANs*, WiMedia Alliance White Paper Online, January 4, 2008. <http://www.wimedia.org/imwp/idms/popups/pop_download.asp?contentID=12443>.

[101] Di Benedetto, M.-G. T. Kaiser, A. F. Molisch, I. Oppermann, C. Politano, and Domenico Porcino, eds., *UWB Communication Systems, A Comprehensive Overview*, Hindawi Publishing Corporation, May 31, 2006.

[102] UWB Forum Information, <http://en.wikipedia.org/wiki/UWB_Forum>.

[103] B. Brackenridge, "Staccato communications." *UWB the WiMedia Way*, Online, May 15, 2007. <http://www.extremetech.com/article2/0,1558,2129883,00.asp>.

[104] S. Wood, *UWB Standards, WiMedia Alliance White Paper*, Online, January 1, 2008. <http://www.wimedia.org/imwp/idms/popups/pop_download.asp?contentID=9042>.

[105] D. Geer, "UWB standardization effort ends in controversy." *IEEE Explore, Computer*, Vol. 39, no. 7, Online, July 2006. <http://ieeexplore.ieee.org/xpls/abs_all.jsp?arnumber=1657900>.

[106] What is standard?, Online <http://www.webopedia.com/TERM/s/standard.htm>.

[107] A. Stone, *What is Wibree?*, Online, January 5, 2007. <http://www.wi-fiplanet.com/news/article.php/3652391>.

[108] J. McCorkle, *Why Such Uproar Over Ultrawideband?*, Online, Mar 01, 2002. <http://www.commsdesign.com/design_corner/OEG20020301S0021>.

[109] Wibree, <http://en.wikipedia.org/wiki/Wibree>.

[110] WiBro, <http://en.wikipedia.org/wiki/WiBro>.

[111] WiMedia Alliance, <www.wimedia.org>.

[112] *WiMedia Alliance Finalizes Certification Program for UWB Common Radio Platform*, Online, December 7, 2006. <http://electronics.ihs.com/news/2006/wimedia-uwb-radio.htm>.

[113] *WiMedia Alliance, MBOA-SIG Merge*, Online, March 7, 2005. <http://www.allbusiness.com/media-telecommunications/media-convergence/6299352-1.html>.

[114] *WiMedia Radio: Current Regulatory Status and Respective Industry Positions*, Aug 2008, Online <http://www.wimedia.org/imwp/idms/popups/pop_download.asp?contentID=13838>.

[115] *WiMedia, The Ultra-Platform for Wireless Multimedia*, Online, 2006. <http://www.wimedia.org/en/events/documents/02WiMedia_Overview_CES2006.ppt>.

[116] A. S. Badesh, *WiMedia UWB PHY Design and Testing*, Agilent Technologies, Online November, 2006. <http://eesof.tm.agilent.com/pdf/badesha_2006_11.pdf>.

[117] WiMAX, <http://en.wikipedia.org/wiki/WiMAX>.

[118] A. Berkema, *WiNet Basics*, Hewlett Packard, Online, Nov 1, 2006. <http://rfdesign.com/next_generation_wireless/transmit_receive_technologies/radio_winet_basics/>.

[119] WiQuest communications, <http://www.wiquest.com>.

[120] *Wireless 1394 Adopts UWB*, Online, Sep 07, 2004. <http://www.unstrung.com/document.asp?doc_id=58893>.

[121] IEEE 802.11-2007 Specification, *Wireless LAN Medium Access Control (MAC) and Physical Layer (PHY) Specifications*, 12 June 2007.

[122] IEEE Std 802.15.1, *Wireless Medium Access Control (MAC) and Physical Layer (PHY) Specifications for Wireless Personal Area Networks (WPANs)*, 14 June 2005.

[123] IEEE Std 802.15.3, *Wireless Medium Access Control (MAC) and Physical Layer (PHY) Specifications for High Rate Wireless Personal Area Networks (WPANs)*, 29 September 2003.

[124] IEEE Std 802.15.3b, *Wireless Medium Access Control (MAC) and Physical Layer (PHY) Specifications for High Rate Wireless Personal Area Networks (WPANs): Amendment 1: MAC Sublayer*, Amendment to IEEE Std 802.15.3, 5 May 2006.

[125] IEEE Std 802.15.4, Wireless Medium Access Control (MAC) and Physical Layer (PHY) Specifications for Low-Rate Wireless Personal Area Networks (WPANs), 8 September 2006.

[126] F. L. Lewis, "Wireless sensor networks," *Smart Environments: Technologies, Protocols, and Applications*. eds. D. J. Cook and S. K. Das, John Wiley, New York, 2004. <http://arri.uta.edu/acs/networks/WirelessSensorNetChap04.pdf>.

[127] C. S. Raghavendra, K. M. Sivalingam, and T. Znati, eds., *Wireless Sensor Networks*, Kluwer Academic, July 1, 2004.

[128] C. Hanudel, *Wireless Universal Serial Bus*, Online <http://www.clarkson.edu/class/cs463/wireless.sp2007/students/WirelessUSB.ppt>.

[129] Wireless Universal Serial Bus Specification, Revision 1.0, USB Implementers Forum, Online, May 2005. <http://www.usb.org/developers/wusb/wusb_2007_0214.zip>.

[130] Wireless USB, <www.usb.org/wusb>.

[131] Wisair, <http://www.wisair.com>.

[132] *Withdrawal of the 802.15.3a PAR*, Online <http://standards.ieee.org/board/nes/projects/802-15-3a.pdf>.

[133] M. Welborn, "XtremeSpectrum CFP presentation." *IEEE 802.15.3-03/153r9*, July 2003. <http://grouper.ieee.org/groups/802/15/pub/2003/Jul03/03153r9P802-15_TG3a-XtremeSpectrum-CFP-Presentation.ppt>.

[134] ZigBee Alliance, <www.zigbee.org>.

Index